X-RAY FLUORESCENCE SPECTROMETRY AND RELATED TECHNIQUES

X-RAY FLUORESCENCE SPECTROMETRY AND RELATED TECHNIQUES

AN INTRODUCTION

EVA MARGUÍ, RENÉ VAN GRIEKEN

MOMENTUM PRESS

MOMENTUM PRESS, LLC, NEW YORK

First published by Momentum Press®, LLC
222 East 46th Street, New York, NY 10017
www.momentumpress.net

ISBN-13: 978-1-60650-391-1 (hard back)
ISBN-10: 1-60650-391-X (hard back)
ISBN-13: 978-1-60650-393-5 (e-book)
ISBN-10: 1-60650-393-6 (e-book)

DOI: 10.5643/9781606503935

Cover design by Jonathan Pennell
Interior design by Exeter Premedia Services Private Ltd.,
Chennai, India

10 9 8 7 6 5 4 3 2 1

Printed in the United States of America

Contents

Preface

Nowadays, X-ray fluorescence spectrometry (XRF) is a well established analytical technique for qualitative and quantitative elemental analysis (usually covering elements of atomic number from Be to U) of a wide variety of samples. In particular, its truly multielement identification capability, acceptable speed, cost-effectiveness, and ease of automation, together with the possibility to directly analyze solid samples are among the most important features of the spectroscopic technique of the many that have made it a very mature analytical tool for routine quality controls in many industries, as well as for analytical support applications in many research laboratories. The recent technological advances, including the design of low-power microfocus tubes and novel X-ray optics and detectors have made it possible to extend XRF to the determination of low-Z elements and to obtain 2D or 3D information at a micrometer size scale. Furthermore, the recent development and commercialization of bench-top and portable instrumentation, which offer extreme simplicity of operation in a low-cost design, have promoted even more the approach of XRF for a wide range of analytical problems in many different and diverse fields.

This tutorial book provides an up-to-date account of the fundamentals of the XRF technique, instrumentation, sample preparation procedures, and applications. It is not focused on one specific type of spectrometer or one application area, but aims to give a broad overview of the main spectrometer types and applications. It is intended for people new to the field of XRF analysis or to help analytical chemistry technicians and researchers and other users of XRF to further expand their knowledge from a practical point of view. For this reason, difficult mathematical equations and X-ray physical postulates have been avoided and only a basic knowledge of these is required. We have tried to keep the language and explanations as straightforward and simple as possible. For more detailed information about XRF technique and applications readers can refer to books listed in the Bibliography of this work.

The first chapter provides a general introduction to the XRF technique including the origin of X-rays and the basic principles. Chapter 2 outlines the basic components of X-ray fluorescence spectrometers and the design of the different XRF systems. Following, in Chapter 3, some considerations for the qualitative and quantitative analysis by XRF are presented. For the latter case, a discussion of matrix effects is discussed and an overview of correction methods is given and commented upon. Information on the sample preparation procedures for various kinds of samples prior to X-ray measurements is reviewed in Chapter 4. Not only are commonly used methods covered, but also some specific techniques for specific kinds of materials are included, with emphasis on more recent applications and developments. Chapters 5, 6, and 7 are focused on the basic information on those XRF-based techniques that are now well established and of recognized value to provide robust and fit-for-purpose analytical results; these include: wavelength dispersive XRF (WDXRF), energy dispersive XRF (EDXRF), and total reflection XRF (TXRF) spectrometry. At the end of each of these chapters a section of selected applications illustrating the possibilities and drawbacks of each particular technique is included. In addition, in Chapter 8, a brief description of special EDXRF configurations and XRF related techniques used in specific fields is presented. Finally, the current role of XRF in chemical analysis and comparison of the analytical capabilities of XRF systems with other elemental techniques such as atomic spectroscopy are discussed in Chapter 9. Trends and future perspectives of XRF are also included in this last chapter. A Buyer's Guide and Reference Section are also included in this book.

KEY WORDS

X-ray fluorescence spectrometry (XRF); Energy dispersive X-ray fluorescence spectrometry (EDXRF); Wavelength dispersive X-ray fluorescence spectrometry (WDXRF); Total reflection X-ray fluorescence spectrometry (TXRF); Micro-XRF; Synchrotron radiation-induced X-ray emission (SRXRF); Metals; Sample preparation; XRF instrumentation; Industrial analysis; Environmental analysis; Geochemical analysis

Series Preface

This *Spectroscopy* series of handbooks is intended to provide a library of books for the many widely used spectroscopic techniques used today in academia, industry, national laboratories, and museums. Each introduces, in a concise manner, the technique's concepts and basic theory and methodology, together with many application examples. Each volume of *Spectroscopy* will enable the reader to gain insight and understanding of the practical implementation of a particular technique, its attributes and limitations, and its current capabilities and developments.

Spectroscopy will cover both atomic and molecular spectroscopy techniques and their wide ranging applications and uses from within the laboratory to out in the field, covering a diverse range of applications, including materials and forensic sciences, security and bio-medical applications, uses within the pharmaceuticals industry, life sciences, and art/archaeological arenas.

Each volume in the series is authored by leading experts on a particular technique. All are individuals recognized for their state-of-the-art knowledge of current developments and practices. This exciting new series was launched by Momentum Press in 2012 with its first volume, *Solid-State NMR: Basic Principles & Practice* (co-authored by David C. Apperley, Robin K. Harris, and Paul Hodgkinson).

This book in the series, co-authored by Eva Marguí and René Van Grieken, is devoted to *X-ray Fluorescence Spectroscopy and Related Techniques*. XRF spectroscopy, although a well-established key atomic spectroscopy technique, continues to evolve to greater sensitivities. This allows the subsequent development of new instruments, which has opened up many new, exciting, and challenging applications including portable instruments for field use.

As the Editor for this *Spectroscopy* series from Momentum Press, I am very pleased to add this excellent XRF spectroscopy book to the series library. Forthcoming volumes in this series include Atomic Absorption Spectrometry (AAS), Laser-Induced Breakdown Spectroscopy (LIBS), and THz spectroscopy.

John M. Chalmers (Series Editor),
January 2013.

Series Editor

John Chalmers, CChem FRSC, has been an independent consultant since 2000, following a distinguished career as a vibrational spectroscopist in the research department measurement science group of a division of a major UK chemical company, the then ICI plc. Mr. Chalmers is a past president of the Society for Applied Spectroscopy (SAS); he served for 9 years as chairman of the UK Infrared and Raman Discussion Group (IRDG); he has served for many years on the Royal Society of Chemistry Molecular Spectroscopy Interest Group. He has held part-time or honorary positions at the Universities of Nottingham, East Anglia and Lancaster. In 1994, Mr. Chalmers was the recipient of the Williams-Wright Award presented by The Coblentz Society for outstanding contributions in the '*Field of Industrial Infrared Spectroscopy*'. He is a co-editor (with Professor Peter Griffiths) of the highly acclaimed 5-volume reference work, *Handbook of Vibrational Spectroscopy*, published in 2002. Additionally, he has co-authored one book and edited or co-edited another 10 books/reference works, authored or co-authored 24 book chapters/reference works, and published over 50 peer-reviewed papers in scientific journals. Among other publishing-related activities, he is the Article Editor for *Spectroscopy Europe*.

About the Authors

Eva Marguí is an Associate Professor in the Department of Chemistry of the University of Girona (Spain). She is a researcher in the Analytical and Environmental Chemistry Unit of the same University. Her research interests are focused on the development of analytical methodologies based on XRF for the determination of elements in environmental and industrial samples. She is the author of more than 30 publications on XRF in high-level journals and has participated in different conferences in the field as an invited lecturer. She is the secretary of the European X-ray Spectrometry Association.

René Van Grieken is a Professor in the Department of Chemistry of the University of Antwerp in Belgium, where he teaches instrumental analysis and environmental chemistry and is the Director of the Environmental Analysis Group within the Micro- and Trace Analysis Center. He is active with major European scientific organizations and has delivered 270 invited lectures at international scientific conferences. He has authored or co-authored over 700 publications with leading scientific journals, book publishers, and institutes in 51 countries. These publications have been cited over 11,000 times.

CHAPTER 1

INTRODUCTION

1.1 BASIC PRINCIPLES OF X-RAY FLUORESCENCE

The principle of the X-ray fluorescence (XRF) technique is based on the use of some excitation radiation to produce ionization in the inner shells of the atoms present in the sample due to photoelectric absorption. The process of atom ionization and X-ray characteristic emission is illustrated in Figure 1.1.

If a photon with energy larger than the binding energy of an inner electron orbital, impinges onto the atom, part of its energy can be absorbed with a certain probability (photoelectric absorption) and the atomic electron is ejected from its atomic position. As long as the vacancy created in the shell exists, the atom is in an unstable state (ionized atom) and there are two processes by which the atom can revert back to its original state. The first process involves a rearrangement that results in the emission of other photoelectrons (Auger effect). The second process, which is related with the origin of X-rays, is based on the transference of an electron from one of the outer orbitals to fill the vacancy. The energy difference between the initial and the final states of the transferred electron is given off in the form of an X-ray photon. Because there are two competing effects by which an atom may return to its initial state, the intensity of an emitted characteristic X-ray beam is dependent on the relative effectiveness of the two processes within a given atom, which is ω, the probability of X-ray characteristic emission occurring (i.e., fluorescence yield).

It is important to note here that the fluorescent yield has a value of around unity for the higher atomic numbers to less than 0.01 for the low atomic number elements. It is mainly for this reason that the sensitivity of the X-ray spectrometric techniques is rather poor for light elements (see Figure 1.2).

The law of Henry Moseley relates the wavelength, λ, of the characteristic emission to the atomic number, Z, as

$$1/\lambda = K(Z-\sigma)^2 \tag{1.1}$$

where K and σ are constants, the values of which are specific for the energy distribution levels and sublevels involved in a given transition. According to the quantum mechanics postulates,

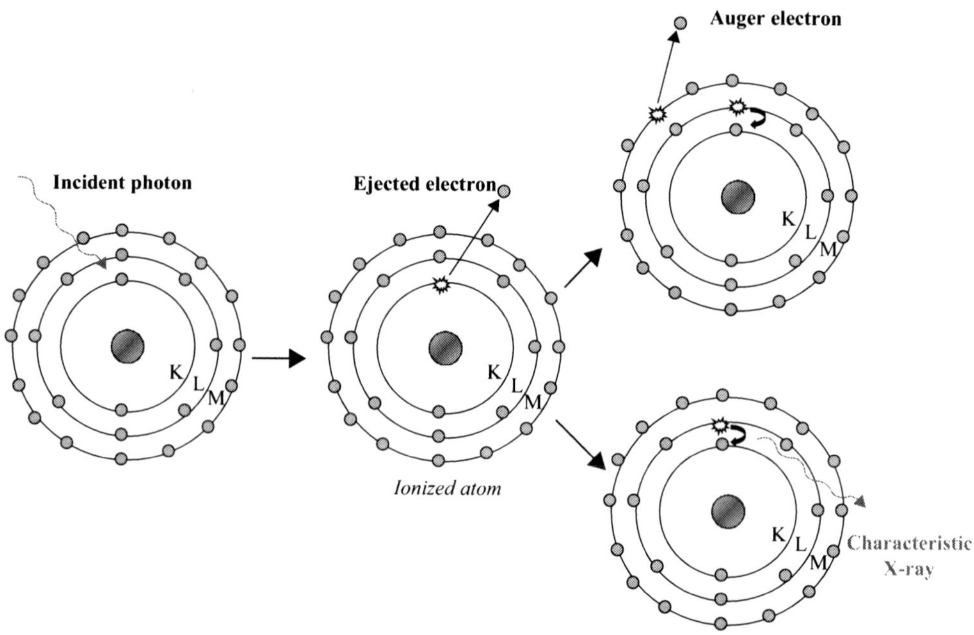

Figure 1.1. Processes of atom ionization and X-ray characteristic emission.

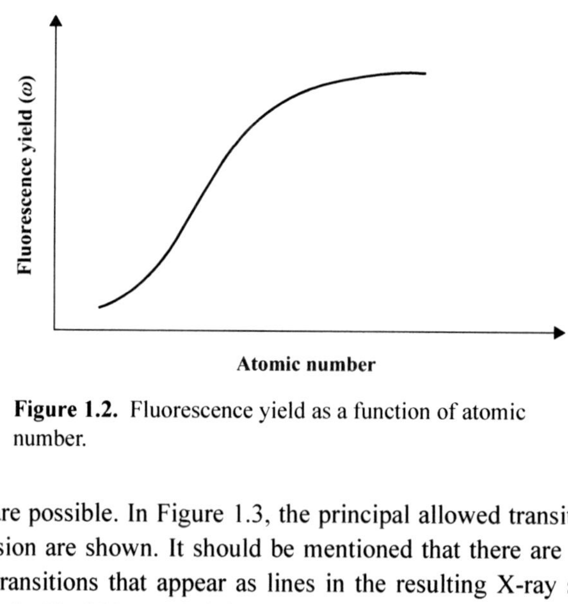

Figure 1.2. Fluorescence yield as a function of atomic number.

not all transitions are possible. In Figure 1.3, the principal allowed transitions yielding X-ray characteristic emission are shown. It should be mentioned that there are several conventions used to name the transitions that appear as lines in the resulting X-ray spectrum. The most commonly used are the Karl Manne Seigbahn (Siegbahn) and International Union of Pure and applied Chemistry (IUPAC) notations. In the first one, the line is named by the symbol of an element followed by the name of the shell in which the initial hole occurred plus a subscript Greek letter (α, β, γ) that indicates the relative intensity of the line. For instance, Zn K_{α} is the strongest zinc line because of an expelled K electron. When using the IUPAC nomenclature, a line is indicated by the symbol of the element, the shell in which the initial hole occurred, and also by the shell from which the electron comes that fills this hole. For example, Zn KL_{III} is zinc radiation due to a hole produced in the K-shell filled by an electron in the L_{III}-shell.

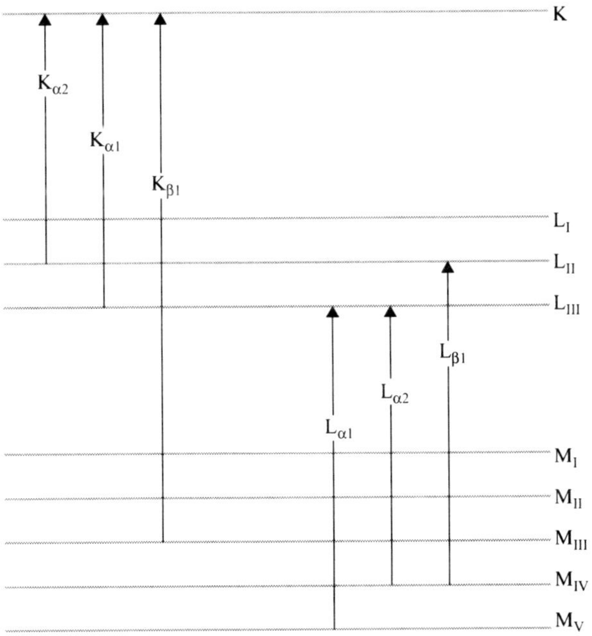

Figure 1.3. Principal allowed transitions producing X-ray characteristic emissions.

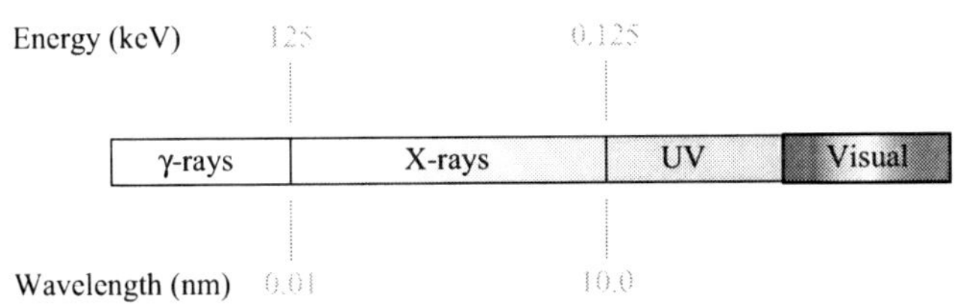

Figure 1.4. Electromagnetic radiation and the X-ray region.

The wavelength, λ, and the energy of the emitted radiation, *E*, are related using the Planck's constant (*h*) and the speed of light (*c*):

$$E = \frac{hc}{\lambda} \tag{1.2}$$

In the specific case of X-ray radiation, associated wavelengths are in the range of 0.01–10 nm, which corresponds to energies in the range from 0.125–125 keV (see Figure 1.4).

To achieve X-ray production from a given element, the only requirement is to use an excitation radiation with an energy larger than the absorption edge for the corresponding group of lines. Most commercially available X-ray spectrometers have a range of about 0.4–20 Å (40–0.6 keV), and this range allows measurement of the K series from F (Z = 9) to Lu (Z = 71), and the L lines from Mn ($Z = 25$) to U ($Z = 92$).

As characteristic emissions correspond to transitions ruled by quantum mechanics principles, the XRF method constitutes a unique technique for the identification of the atoms present

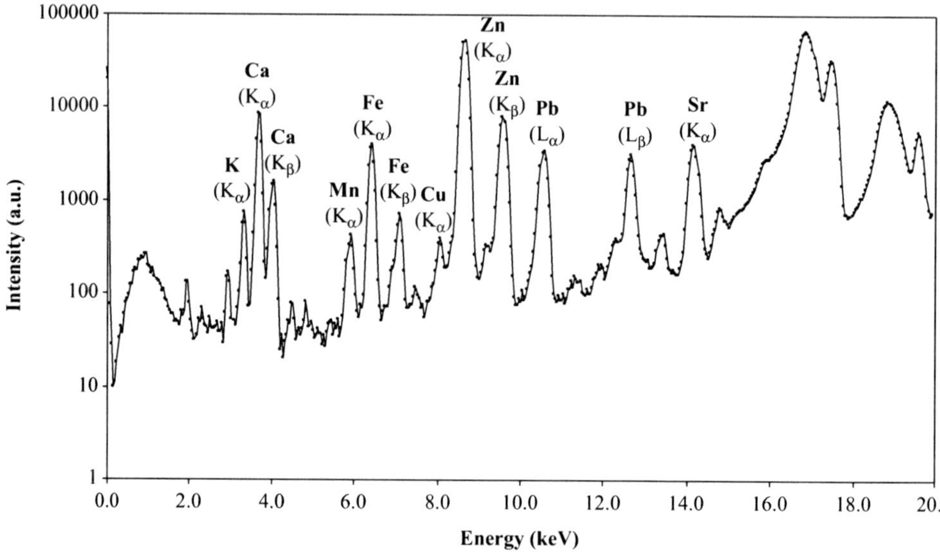

Figure 1.5. Typical spectrum of a vegetation sample measured with an XRF spectrometer.

in a sample (i.e., qualitative analysis). A straightforward identification of any element present in the sample is possible by detecting the presence of several of its characteristic emission energies (or wavelengths). As the X-ray production is (almost) independent of the chemical association of the elements (i.e., transitions arise from inner orbitals), the method also has a multielemental capability. As an example, in Figure 1.5, a typical X-ray spectrum for the analysis of a vegetation sample is shown. The positions of the characteristic X-ray emitted energies determine the elements present in the sample, while the heights of the peaks are related to the element concentrations.

1.2 INTERACTIONS OF X-RAYS WITH MATTER

In the previous section, we considered that the unique phenomenon resulting in the emission of electromagnetic radiation that takes places when a beam of X-ray photons falls onto a given specimen, was fluorescence. However, when the incident X-ray photon beam impinges on the sample, a fraction of this radiation can also be transmitted through the sample or scattered back from it, as is illustrated in Figure 1.6.

Scatter occurs when an X-ray photon interacts with the electrons of the target element. When this interaction is elastic (i.e., no energy is lost in the collision process), the scatter is named as *coherent* (Rayleigh) *scatter*. As no energy change is involved in this process, the Rayleigh radiation will retain exactly the same energy as the incident beam. However, the scattered photon can also transfer a part of its energy during the collision. In this case, it is referred to as *incoherent* (Compton) *scatter* and this radiation is characterized by a slightly lower energy compared to that from the incoming beam. Because of these phenomena, in addition to the characteristic radiation of the elements present in the irradiated sample, in an XRF spectrum one finds peaks that correspond to a Rayleigh and Compton scatter of the incident beam. In Figure 1.7, the appearance of Rayleigh and Compton peaks in an XRF spectrum

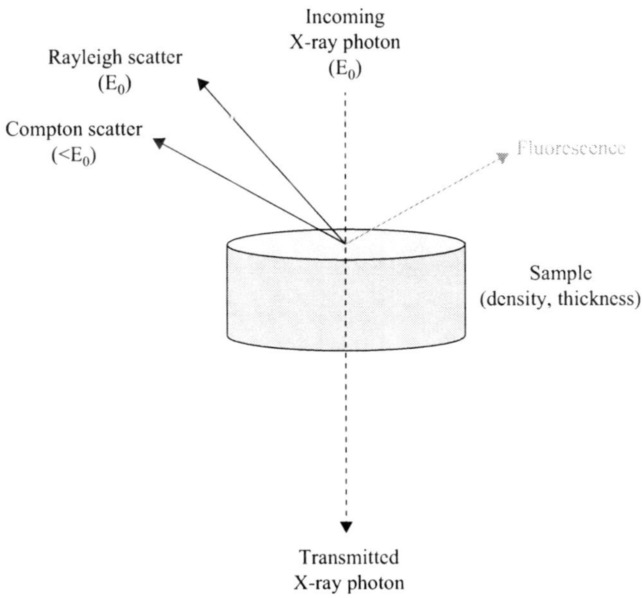

Figure 1.6. Main interactions of X-rays with matter: fluorescence, transmission, and scatter.

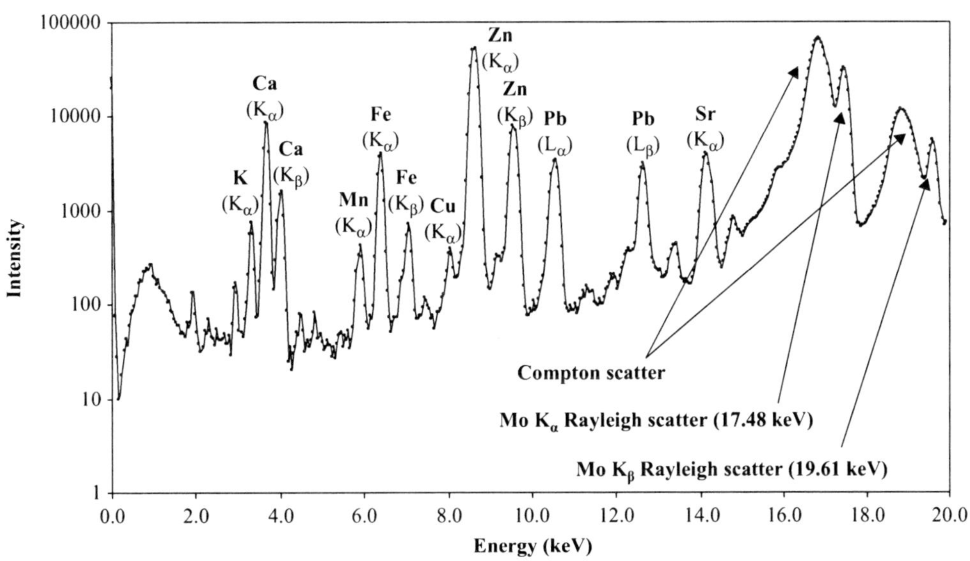

Figure 1.7. Appearance of Rayleigh and Compton peaks in an XRF spectrum obtained when using a Mo X-ray source.

obtained when using an incident Mo X-ray source are displayed. As is shown, Rayleigh scatter peaks are characterized by sharp shapes that correspond to the characteristic energies of the incident beam (in this case K_α and K_β of Mo). The spread of energy in the Compton scatter is larger than for elastic scatter and in the X-ray spectrum this can be observed by the Compton peaks being wider than the Rayleigh peaks and located at slightly lower energies.

From a practical point of view it is interesting to note, that the intensity of the scatter is dependent on the average atomic number, the density and the thickness of the sample being irradiated. For instance, samples with light elements give rise to high Compton scatter and low Rayleigh scatter.

1.3 X-RAY SAFETY AND PROTECTION

As already stated in the previous sections, X-ray radiation has enough energy to eject electrons from atoms, and for that it is classified as an *ionizing radiation.* From the point of view of human health and safety, ionizing radiation is of great concern as it can penetrate deep into the human body, strip electrons from their orbits, and thus break or modify chemical bonds in critical biological molecules that make up the cells. This process can cause cell injury and even cell death. For instance, if enough cells in an organ are injured or altered, the functioning of the organ can be impaired.

Obviously, several factors contribute to the effects of the human body's exposure to X-ray, including:

- dosage (H, expressed in Sievert units, Sv) and total dose received;
- energy of the radiation;
- area of the body exposed;
- individual and cell sensitivity.

Therefore, to avoid any adverse effect to human health and to ensure safe operation of X-ray devices, protective measures that restrict or control access to X-ray areas and devices or warn of X-ray hazards are essential. In all countries, the dose limits are prescribed by law and all values are based on recommendations of the International Commission on Radiological Protection (ICRP).[1] For all European Member States, the Council Directive 96/29/Euratom[2] is the basis for any national regulations. In Table 1.1, dose limits for exposed workers according to this European Union (EU) Directive are shown.

Table 1.1. Dose limits for exposed workers according to the EU Directive 96/29

Organ	Annual limit for exposed workers (in mSv)
Whole body	20
Lenses of the eye	150
Hands, forearms, feet, and ankles	500
1 cm^2 of the skin	500

[1] ICRP Website (www.icrp.org/index.asp); accessed October 2012.

[2] European Commission, Council Directive 96/29/Euratom laying down basic safety standards for the protection of the health of workers and the general public against the dangers arising from ionizing radiation, OJ L 159, 29.6.1996; Bull.5-1996 (1996).

Prior to authorization by the competent authorities of an X-ray facility, a system of radiation protection is required which includes, for EU member states, among other things:

- identification of the radiological risk (for every workplace);
- implementation of radiation monitoring instruments;
- medical surveillance;
- document, archived information on training of workers;
- qualified experts on location.

With regards to the implementation of radiation monitoring instruments, in most countries, dosimetry services are available for individual monitoring (whole body or finger ring dosimeters). Dosimeters are issued at regular intervals which process an individual's dose value during the target monitoring period. These dose values are stored for several years and summed to calculate the annual dose value, which should not exceed the dose limits according to the national legislation.

It is also important to remark that commercial analytical X-ray instruments are shielded to avoid the human body becoming exposed to the beam during normal operation. Special attention has to be given when working with open-beam systems (e.g., those usually employed in the research field) as one or more X-ray beams may not be enclosed.

In summary, readers interested in the field of XRF analysis have to be particularly aware of the risks to the human body from ionization sources and they must know proper radiation protection practices while working with or around X-ray generating devices.

CHAPTER 2

Basic Components of X-Ray Fluorescence Spectrometers

2.1 GENERAL INTRODUCTION

Considering the fundamental principles of X-rays discussed in the previous chapter, the basic components of any XRF spectrometer are:

- an excitation source that produces ionization owing to photoelectric absorption of the atoms present in a sample;
- a specimen presentation system;
- a detection system that collects the characteristic radiation emerging from the sample;
- a data collection and signal processing system.

In addition, some XRF systems are also equipped with source or detector modifier devices to improve the analytical performance for specific applications. In Figure 2.1, the basic components of XRF spectrometers are shown.

In this chapter, the main types of sources, modifiers, and detector systems will be considered. Depending on the type and spatial configuration of these components, different XRF set-ups are, at present, commercially available. These will be discussed in Section 2.6 of this chapter.

2.2 EXCITATION SOURCES

2.2.1 X-RAY TUBES

By far, the most common radiation sources used in X-ray spectrometry are X-ray tubes. The basic design of an X-ray tube is shown in Figure 2.2.

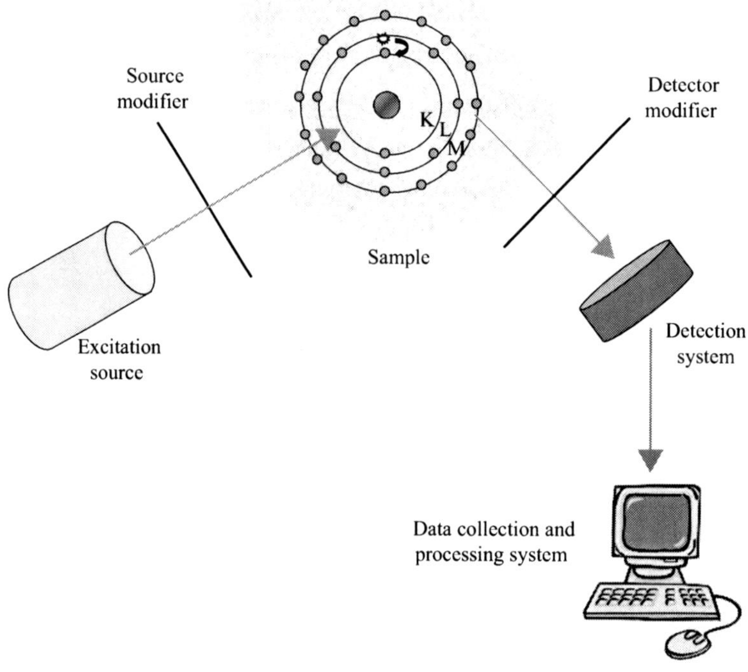

Figure 2.1. Basic components of an X-ray spectrometer.

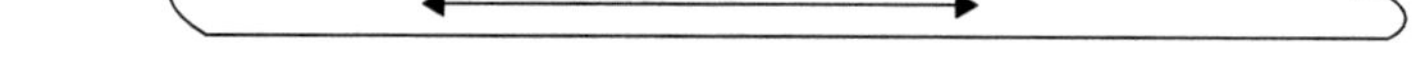

Figure 2.2. Basic design of a side-window X-ray tube (Courtesy of PANalytical).

X-ray tubes contain a spiral filament, usually made of W, acting as the cathode and an anode (usually made of Mo, Rh, Pd, or W) placed inside a vacuum housing. When the filament is fed with a heating current (1 mA–1 A), it emits electrons that are attracted and accelerated in the direction of the anode by means of the application of a high voltage (30–150 kV) between the cathode and the anode. When the electrons hit the anode they are decelerated, which causes the emission of X-ray photons. This radiation is generally called the "Bremsstrahlung" (from the German *brems* for braking and *strahlung* for radiation); it appears in the XRF spectrum as a continuous band of emitted energies. In addition, a fraction of the electrons that impinge onto the atoms of the anode will eject electrons from these atoms producing emission of characteristic radiation, as explained in Chapter 1 (see Figure 2.3). The energy of the characteristic radiation of the X-ray tube is determined by the element from which the anode is made while the continuous radiation depends on the applied voltage (kV), the current (mA), and on the anode material used.

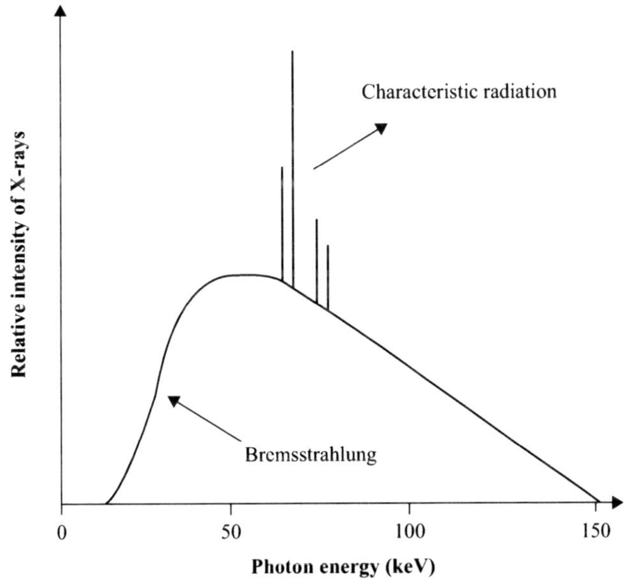

Figure 2.3. Emission from an X-ray tube.

Both types of emerging radiation are emitted in all directions but they emerge from the X-ray tube only through a Be window that it is placed at the end (end-window X-ray tube) or to one side (side-window X-ray tube) of the X-ray tube.

In addition to X-ray emission, some energy is released in the form of heat by the anode material and therefore, the X-ray tube must be water cooled to avoid any damage. For this reason, XRF instrumentation is usually equipped with an external water refrigeration system with water consumption typically of about 4 l/min at a pressure of 3–5 bar and a temperature of 20°C–25°C. Although, recent technological advances have led to the design of very compact low-power metal-ceramic X-ray tubes (with a maximum power of 50 W) that can be cooled by air and therefore, no external cooling media is required. This has enabled development of benchtop and handheld systems which may be purchased at a very modest instrumental cost investment.

An X-ray tube's lifetime can span some 3000–6000 working hours (mostly it depends on the anode material). To ensure a long life, operation warm-up conditioning should take between 15 and 45 min depending on the previous period of interruption, and the cooling down time should last for at least 10 min. In addition, the manufacturer's recommended value for maximum power of operation should never be exceeded.

So far, we have described the basic principles and operation of X-ray tubes. From an analytical point of view, the implications for the selection of an X-ray tube for XRF analysis are:

- **The anode material of the X-ray tube and the voltage applied (kV) determine which elements can be excited:** The atoms of the elements present in the sample can be excited by means of the continuum radiation of the X-ray tube and also by the characteristic energies emitted by the X-ray anode material (K and L lines). In general, most of the excitation of the low-energy characteristic lines of the sample arise from the low-energy characteristic lines of the tube and most of the higher-energy excitation from the continuum radiation from the tube. In Figure 2.4, an example of an XRF spectrum obtained from the analysis of a fly ash sample using a Rh anode X-ray tube is shown.

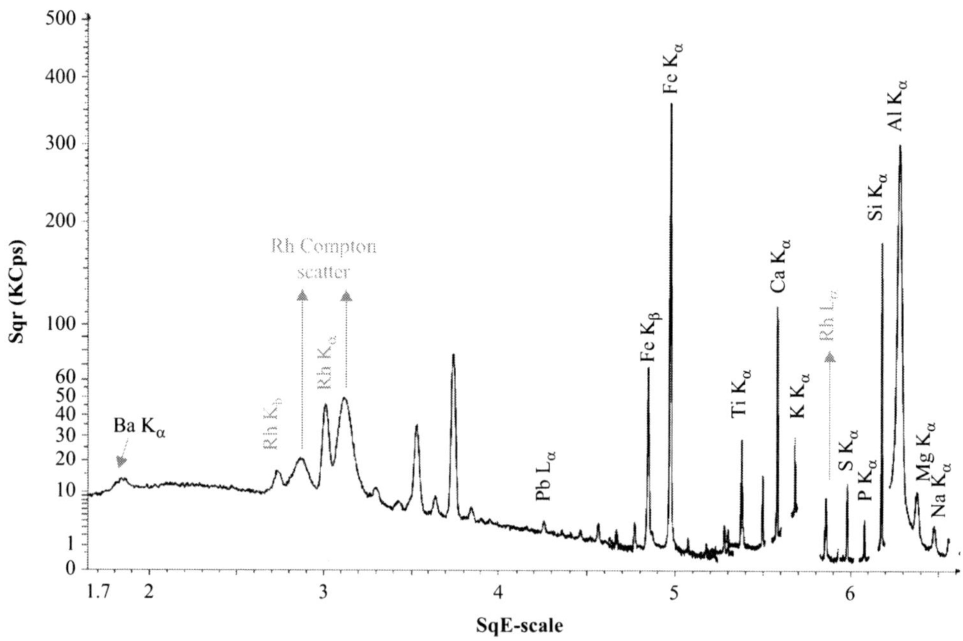

Figure 2.4. XRF spectrum obtained in the analysis of a fly ash sample using a Rh anode X-ray tube (Wavelength dispersive XRF system).

As can be seen in Figure 2.4, Rh L lines are effective for exciting the light elements in the range 1–2.5 keV ($Z < 16$). If higher energy lines (e.g., Fe K_{α} at 6.4 keV) are sought, the Rh L lines provide no excitation as they are lower in energy than the absorption edge (FeK at 7.11 keV) and the Rh K lines then provide the necessary excitation. Elements with energies higher than Rh K lines (such as Ba K_{α}) can be determined by means of the continuum radiation or by using a secondary X-ray line (L lines) and using as excitation energy, the Rh K line. According to this, and depending on the anode material and the voltage applied to the X-ray tube, one will be able to excite one particular range of elements or another. For instance, in Figure 2.5, a periodic table is displayed of the elements that are usually excited using Rh X-ray tubes along the characteristic X-ray lines used for measurement of each element in the periodic table.

Therefore, depending on the material of the X-ray tube, one will be able to detect a predefined range of elements with a higher or lower sensitivity depending on the measurement line (K or L) used for each analyte. To extend the range of excited elements without losing analytical sensitivity, some XRF systems are equipped with a combination of two different X-ray tubes. Usually one of the tubes is used for the excitation of lighter elements (e.g., Sc, Cr) and the other one for the determination of heavier elements (e.g., Mo, W).

- **The power of the X-ray tube is related to the sensitivity of the instrument:** The flow of electrons between the cathode and the anode is dependent on the current applied to the X-ray tube (expressed in mA) while the speed at which the electrons travel and strike the anode depends on the applied voltage (expressed in kilovolts). Therefore, both parameters influence the intensity of the X-ray radiation emerging from the tube, as shown in Figure 2.6. Taking this into account, the power of the X-ray tube, which is expressed as the product of the current and the high voltage, will be one of the most important parameters related to the sensitivity of XRF systems.

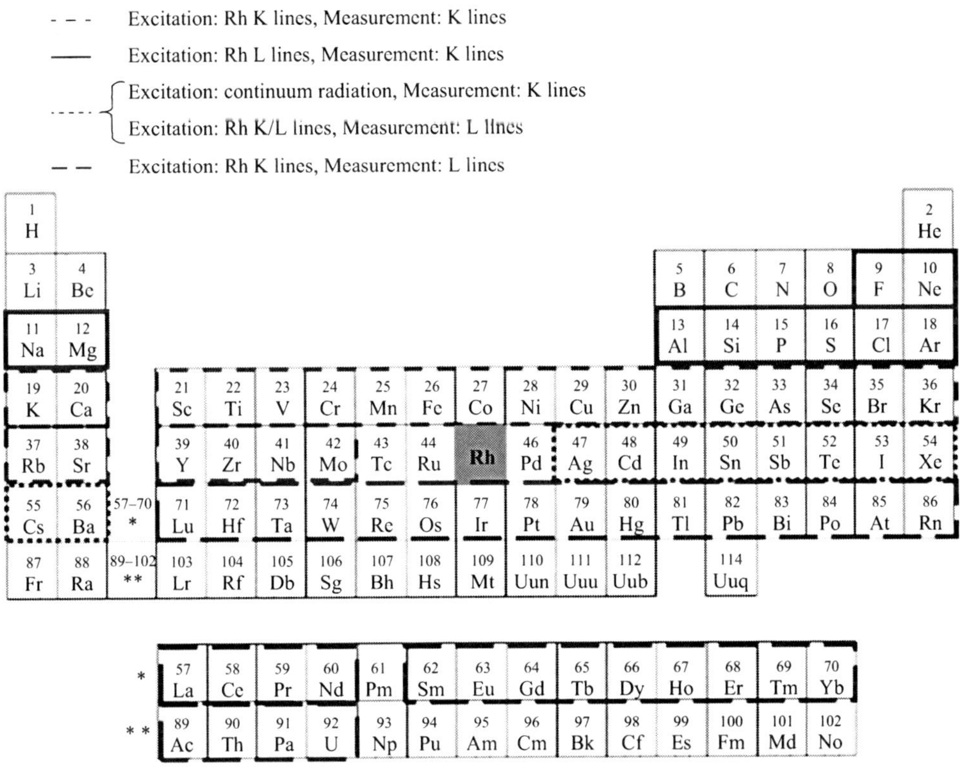

Figure 2.5. Elements that are usually excited using a Rh X-ray tube and the characteristic X-ray lines used for element measurement.

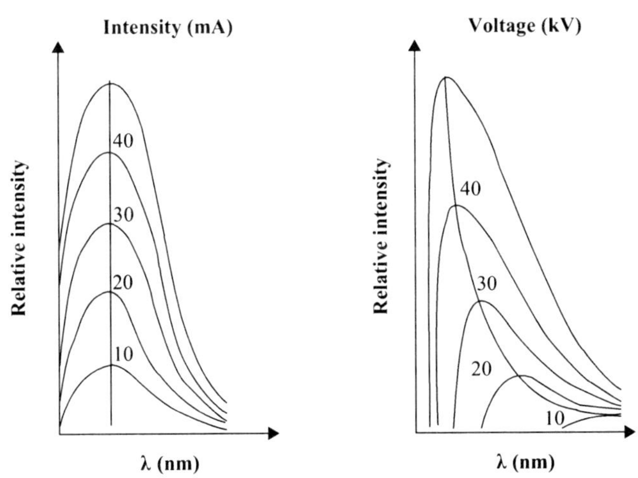

Figure 2.6. Influence of the applied voltage (kV) and current (mA) on the X-ray radiation emerging from the tube.

In Table 2.1, maximum power values associated with the most common commercial X-ray tubes are presented. Usually, low-power X-ray tubes (50 W) present a lower sensitivity compared to high-power X-rays tube of 1–4 kW. To increase the analytical performance of the former systems, source modifiers are introduced between the source and the sample to improve the signal-to-noise ratio and the instrumental limits of detection. In Section 2.2.2, we will discuss in more detail the types and benefits of such devices.

Table 2.1. Maximum power values of most commercial X-ray tubes

Conventional X-ray tubes		
Maximum power (W)	1000	4000
Voltage (kV)	20–50	20–60
Intensity (mA)	50–20	5–150
Air cooled X-ray tubes (low-power X-ray tubes)		
Maximum power (W)	50	
Voltage (kV)	10–50	
Intensity (mA)	0.1–1	

Table 2.2. Most commonly used radioisotope sources

Isotope	**^{55}Fe**	**^{109}Cd**	**^{241}Am**	**^{57}Co**
Energy (keV)	5.9	22.8	59.5	122
Elements (K-lines)	Al–V	Fe–Mo	Ru–Er	Ba–U
Elements (L-lines)	Br–I	Yb–Pu	None	None

2.2.2 *RADIOISOTOPES*

Radioactive X-ray sources (such as ^{55}Fe, ^{109}Cd, ^{241}Am) were very popular in field-portable XRF analyzers until very recently. Their small physical size (small cylinders of approximately 5 mm height and diameter) and their zero power consumption supported their utilization in portable and compact devices. However, certain drawbacks associated with these X-ray sources are the rather short half-lives of some isotopes (e.g., ^{55}Fe, 2.7 years; ^{109}Cd, 453 days), their low intensity, and the need, in most of the countries, of specific regularity requirements for their transportation and field applications.

In Table 2.2, the most commonly used radioisotope sources with their associated X-ray energy are displayed.

In spite of the recent decrease in the use of radioisotope sources owing to the development of low-power metal-ceramic X-ray tubes, they are still used, even today, in some specific XRF applications. For example, in screening for RoHS (Restriction of Hazardous Substances Directive) in electronic devices and components, the ^{57}Co, ^{241}Am, and ^{109}Cd sources have been adopted by some manufacturers, in particular, the first two that do not have a direct equivalent with other available portable X-ray sources.

2.2.3 *OTHER SOURCES*

In addition to the commonly used X-ray tubes and to a lesser extent the radioisotopes, there are other sources for X-ray analysis based on the use of accelerated particles. Among them, the most important are the use of synchrotron radiation (electromagnetic radiation from decelerating high-energy electron beams), particle induced X-ray emission (from proton or alpha particle beams), and electron induced X-ray emission (from electron beams in scanning electron microscopes or microprobes). However, these sources (in particular the former two) are not commonly available to the general user community, as in most cases large beam national facilities (accelerators) are needed. For this reason, only the main principles, benefits, and particular fields of application of these sources will be summarized later in Chapter 7, which focuses on special XRF configurations and related techniques.

Taking into account that the most commonly used excitation sources for X-ray analysis (especially in commercial systems) are X-ray tubes, in the following sections we will focus on XRF systems using X-ray tubes as excitation sources.

2.3 SAMPLE CHAMBER

The specimen presentation system is built as an integral part of the spectrometer itself. It is located between the X-ray source and the detector system. Usually, it consists of a sample chamber in which the test sample to be measured is placed. To increase throughput, some systems are equipped with a flexible *x–y* sample changer with capacity for more than 100 samples (see Figure 2.7(a)). The samples, placed within circular stainless-steel cups, are sequentially introduced into the sample chamber by means of a mechanical arm system.

In some XRF instruments, the sample chamber can operate under vacuum conditions. This condition is crucial if the determination of light elements (low energies) is required, as the characteristic X-rays from these elements can be absorbed into air and consequently not reach the detector. On the other hand, liquids and wet powders cannot be measured in vacuum conditions because they may evaporate. In such cases, the sample chamber is filled with a helium flow. Helium absorbs the radiation of the light elements (up to F) but it does not affect the radiation from heavier elements.

Frequently, in systems that operate in open-air conditions, it is not necessary to place the sample to be analyzed in a cup of fixed external dimensions in a radiation protection chamber (see Figure 2.7(b)). Therefore, these systems have no constraints with respect to the dimensions and shape of samples.

As mentioned earlier, samples are usually introduced into stainless-steel cups to perform the measurements. However, other sample holders can be used in specific applications.

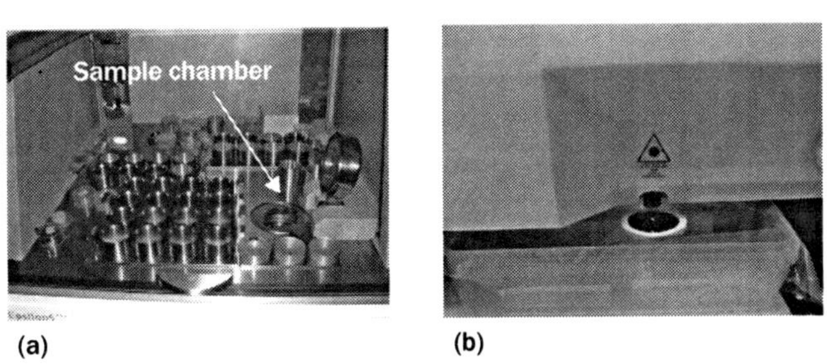

Figure 2.7. Specimen presentation systems: (a) Flexible *x–y* sample changer, (b) System working in air conditions.

Finally, it is important to remark that samples are not always perfectly homogeneous, and scratches on the surface can influence the analytical results. To overcome this limitation, in some XRF instruments, a device called a *spinner* rotates the sample during the measurement to reduce the influence of these detrimental effects.

2.4 DETECTION SYSTEM

The aim of any XRF spectrometer detection system is to convert X-ray photons emerging from the sample into electrical pulses (counts per second). The height of this pulse is proportional to the energy of the incoming photon. The pulses are then amplified and counted by a multi channel analyzer (MCA). The number of pulses of a certain height gives the intensity of the corresponding energy and it is shown as a peak in the XRF spectrum. The natural width of fluorescence lines is normally less than 10 eV and varies insignificantly with the type of lines series (K, L, M) and the chemical bond type of the atom. However, because of thermal detector noise and subsequent amplification electronics, peaks in a digitalized XRF spectrum are not displayed as sharp lines but spread by a Gaussian-type curve (Figure 2.8).

2.4.1 TYPES OF DETECTORS

At the time of writing this book, there are three main types of detector systems in the market which are used in X-ray analysis: gas proportional counters, scintillation detectors, and semiconductor detectors. In this section, the basic principles and mode of operation of these detectors are summarized.

2.4.1.1 Gas-filled Detectors

Gas-filled detectors consist of a metal cylindrical tube (often of Al) of about 2 cm in diameter, carrying a thin (25–50 μm) W wire along its radial axis which is raised to a high voltage (1300–2000 V). A Be window allows X-ray photons to enter the detector that is filled with an inert counting gas (e.g., Ne, Ar, Kr, Xe, or He). A schematic of this type of detector is shown in Figure 2.9.

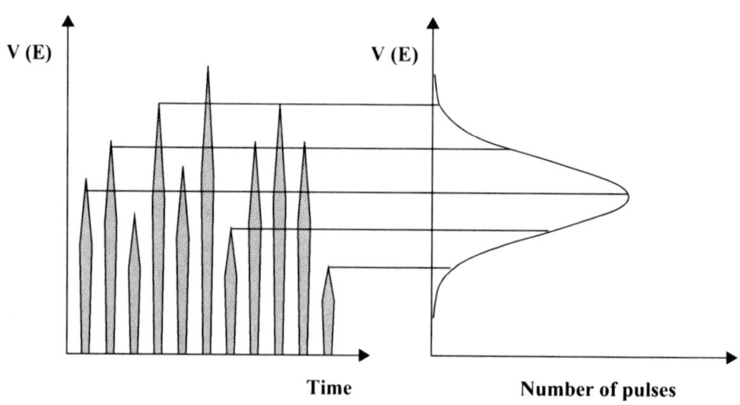

Figure 2.8. Conversion of the electric pulses (a) into XRF spectrum peaks (b).

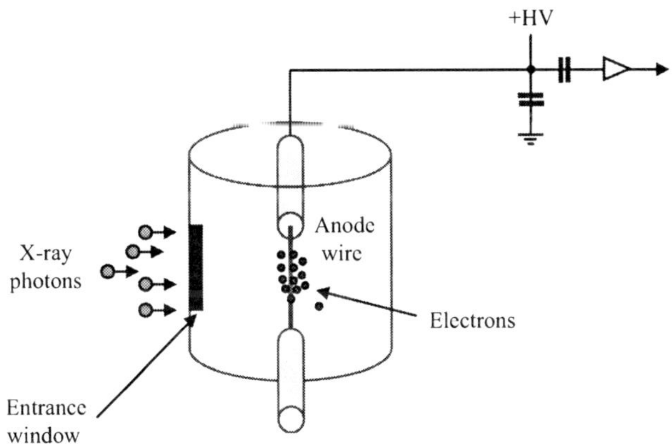

Figure 2.9 Graphic scheme of a gas-filled detector. HV, high voltage. (Courtesy of PANalytical).

The operation of a gas-filled detector involves the generation of a small cloud of electrons when the X-ray photons enter the gas-filled detector. The electrons then reach the anode wire and cause a drop in the voltage which is registered as a negative pulse in the amplifier. The number of electrons is proportional to the energy of the incoming radiation and thus the height of the pulse. A MCA counts the pulses produced by the detector.

2.4.1.2 Scintillation Detectors

Scintillation detectors consist of two main parts: the scintillation crystal and the photomultiplier. X-ray photons pass through the Be window and are incident on the scintillation crystal (usually a large single crystal of NaI doped with Tl) which produces a blue-light flash. These blue-light photons, which are related to the energy of the incident X-ray beam, produce electrons by interaction with a photosensitive surface in the photomultiplier, and the number of electrons is linearly increased by a series of secondary surfaces called dynodes. When the resulting electrons reach the anode, they cause a drop in the voltage, as for the case of the gas-filled detectors. This is registered as a negative voltage pulse in the amplifier. As the number of electrons is proportional to the energy of the incident X-ray photon, the scintillation counter is a proportional counter. In Figure 2.10, a schematic of a scintillation detector is shown.

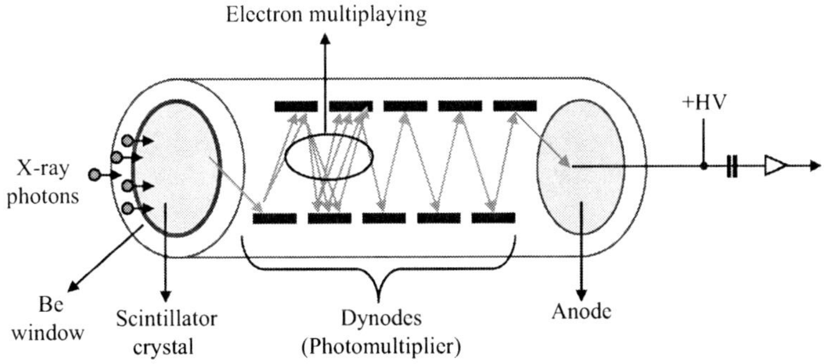

Figure 2.10. Graphic scheme of a scintillation detector. HV, high voltage. (Courtesy of PANalytical).

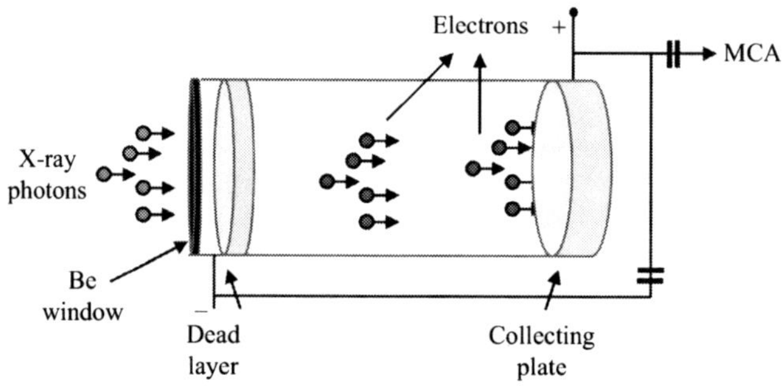

Figure 2.11. Graphic scheme of a solid-state detector. MCA, multichannel analyzer. (Courtesy of PANalytical).

2.4.1.3 Solid-state Detectors

Solid-state detectors are constructed with a body of a semiconductor, usually lithium-drifted silicon (Si(Li)) or high-purity germanium (HPGe), but compound semiconductors such as HgI_2, CdTe, CdZnTe, or GaAs have also been used to varying degrees. When an X-ray photon is incident upon the semiconductor (dead layer), it normally interacts by photoelectric absorption to create electron–hole pairs in the semiconductor material. Then the created electrons are attracted to the back of the dead layer by the application of a high voltage (1500 V) between the dead layer and the back. When the electrons reach the collecting plate, the potential drops and gives a negative pulse. The intensity of the pulse is proportional to the number of electrons and thus, proportional to the energy of the incoming radiation. In Figure 2.11, a schematic of a solid-state detector is shown.

It is essential that the Si(Li) detector is kept cold throughout the time that the high voltage is applied. If this is not done, then there is the possibility of damage to the detector assembly. For this reason, liquid nitrogen cryostats are required when operating this kind of detector.

In addition to the conventional Si(Li) detector, Si(PIN) diodes and Si drift detectors (SDD) have also become available in recent years. Like other solid-state X-ray detectors, Si(PIN) photodiodes and SDDs measure the energy of an incoming photon by the amount of ionization it produces in the detector material. The first type consists of a semiconductor diode with three layers. The usual P and N regions are present, but between them there is a layer of intrinsic material with no doping, and as a result this increases the size of the depletion region compared to the Si(Li) detectors. In the SDD, the detector material is high purity Si with a very low leakage current. The high purity allows the use of thermoelectrical cooling (Peltier effect) instead of the traditional liquid nitrogen cooling. This is one of the main advantages of these detectors, above all, for their convenience in the design of field-portable instrumentation.

2.4.2 RESOLUTION AND EFFICIENCY

From an analytical point of view, the two most important properties of any detection system used in X-ray analysis are: energy resolution and detection efficiency.

2.4.2.1 Resolution

Resolution is defined as the ability of the detector to distinguish between different energy levels. For instance, in Figure 2.12, XRF spectra obtained using detector systems with different resolutions are displayed. As can be seen, when using a resolution of about 220 eV, we are not able to distinguish between the Al and the Si peak. The same trend is observed for K and Ca peaks. However, when using a system with a higher resolution (~140 eV), we are able to distinguish between the pairs of peaks in both cases.

Usually the resolution is estimated by the Full Width at Half Maximum (FWHM) which defines the peak width at the half peak maximum as is shown in Figure 2.13.

To cope with the fact that FWHM values increase with energy and to compare the performance of X-ray detector systems, it has been agreed to use the Mn-K_α as a reference line of the FWHM.

Resolution is an important parameter to be considered when developing an XRF method because it will influence the selectivity of the methodology. This is especially the case for

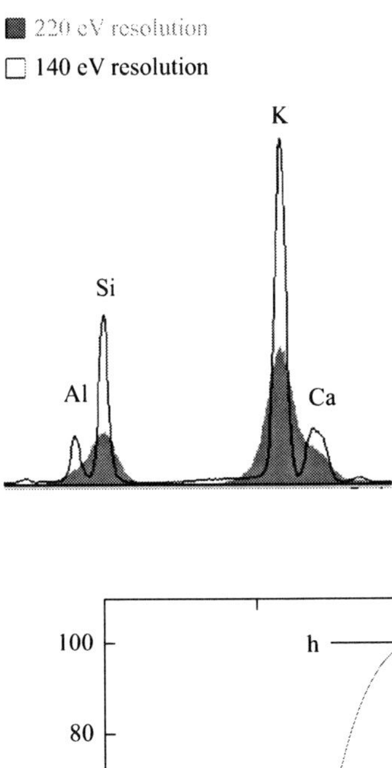

Figure 2.12. Graphical view of XRF spectra obtained using a detection system with different resolution.

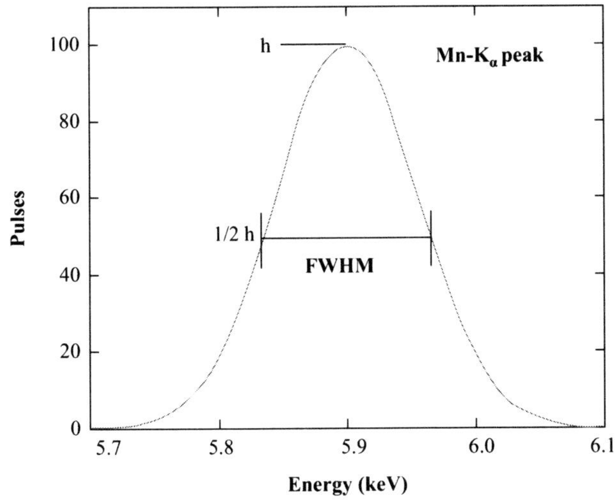

Figure 2.13. Definition of the resolution. FWHM, Full Width at Half Maximum.

multielement samples; strong overlaps of the individual fluorescence peaks in the spectrum occur because of the line diversity and the finite detector resolution. To allocate the correct intensities to the individual elements, the spectrum is usually separated into the individual element lines by a complex mathematical process (deconvolution). We will discuss this procedure in more detail in Chapter 3.

2.4.2.2 Efficiency

Detector efficiency is defined as the percentage of ionizing radiation reaching the detector that is measured. Therefore, detector efficiency is high if the ratio of the number of pulses against the number of incoming photons is high.

The detector efficiency depends on different parameters including the type of detector, the detector size and shape, the distance from the detector to the sample, the absorption of radiation before it reaches the detector (by air and by the detector covering), and the backscatter of radiation toward the detector. In Figure 2.14, the effect of detector type and thickness of the Be window of a detector are compared.

Usually, poor efficiency at low energies is related to the combined absorption at the Be entrance window and the air path between sample and detector. The plot also shows the high efficiency for both detectors (Ge and Si(Li)) over a wide range of useful X-ray energies. The Ge ($Z = 32$) detector is efficient at much higher energies than a Si ($Z = 14$) detector because of the higher atomic number and subsequent larger photoelectric probability.

2.4.3 COMPARISON OF DETECTION SYSTEMS

The selection of a detector system is usually based on the resolution and efficiency requirements for the elements to be determined. In Table 2.3, some characteristics and usual applications of common X-ray detectors are summarized.

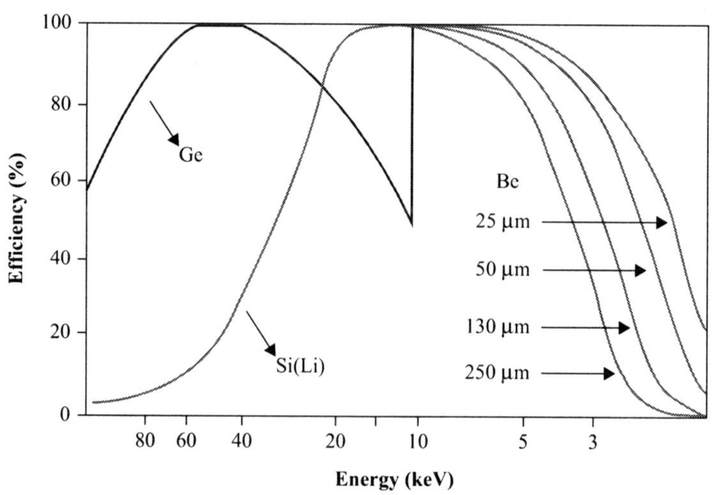

Figure 2.14. Effect of detector material and Be detector window thickness on the detector efficiency.

Table 2.3. Characteristics of common X-ray detectors

Detector type	Efficiency	Resolution at Mn-K_α (eV)	Count rate (cs)	Cooling media
Gas-filled detector	Good for light elements (Be–Cu)	500–1000	10,000–1,000,000	Room T
Scintillation detector	Good for medium–high elements (Cu–U)	>1000	10,000–1,000,000	Room T
Solid state detectors				
Si (Li)	Wide-range efficiency (Na–U)	135–140	3000–50,000	Liquid N_2
HPGe		<140	3000–50,000	Liquid N_2
PIN diode		150–160	3000–20,000	Peltier effect
SDD		<140	10,000–300,000	Peltier effect

As shown in Table 2.3, gas proportional counters are X-ray detectors characterized by moderate energy resolution, room-temperature operation, and high counting rates capability. Their efficiency is good for light elements (Be–Cu) but it is limited for medium–high atomic number elements (Cu–U). In such cases, scintillation detectors are a good alternative. In comparison to the gas detectors, they are based on a higher Z material with a higher density and often they have a greater thickness. This means that they have a high probability of detecting photons over a wider X-ray energy range (Cu–U). However, the energy resolution offered by scintillation detectors is also very poor compared to solid-state detectors, especially in the low-energy range. For this reason, gas-filled and scintillation detectors are usually used (as we will explain in more detail in Chapter 5) exclusively in wavelength dispersive XRF (WDXRF) systems. In these systems, owing to the high resolution power of single crystals or multilayered structures introduced between the sample and the detector system (resolution ~15–30 eV), photons corresponding to characteristic lines with close energies can be detected without interfering with each other. Therefore, as spectral interference is avoided, the detection can be performed using gas-filled or scintillation detectors with a poor resolution but with large efficiency at even low energies.

Unlike the WDXRF systems, the conventional Energy Dispersive XRF (EDXRF) spectrometers consist of only two basic units: the excitation source and the spectrometer-detection system. For this case, as the resolution of the energy dispersive system is equated directly to the resolution of the detector, typically solid-state detectors of high intrinsic resolution are employed. The Si(Li) detector is currently the most popular X-ray detector in the energy range of few hundred electron volts up to 40 keV. The typical energy resolution of commercial detectors is of the order of 135 eV at Mn-K_α. HPGe can also be used for high-resolution X-ray detection. Unlike Si(Li) detectors, Ge detectors do not need to be cooled permanently, but only when they are in operation. Compared with Si, Ge is characterized by a better efficiency for higher energy photons because of the higher atomic number and density of Ge than Si.

The breakthrough in X-ray instrumentation has been, of course, the development of the new generation of compact sized solid-state X-ray detectors, based on thermoelectrical cooling (Peltier effect) detectors (PIN diodes and SDDs), which exhibit energy resolution and signal-to-background analytical performance competitive to the conventional liquid nitrogen cooled Si(Li) detectors. An historical step in ED X-ray detection was the passing of the 200 eV FWHM barrier in the mid-1990s using a Si(PIN) diode X-ray detector. Today, Si(PIN) diodes are commercially available with a typical crystal thickness of 0.5 mm, an active area of 5–25 mm^2 and an FWHM at 5.9 keV down to 150 eV. In recent years, much attention has been given to the SDD. These detectors can be operated at much shorter shaping times than a Si(PIN), with comparable resolution, whereas under optimum conditions the SDD resolution is significantly better than the Si(PIN). Furthermore, the main feature of an SDD is the possibility to record high count rates, up to 300,000 cps, without significant deterioration of its energy resolution. Typical characteristic of SDD detectors are that their FWHM at 5.9 keV is below 140 eV, with a crystal thickness of 0.5 mm and active area dimensions up to 80 mm^2. Owing to the possibility of using PIN diodes and SDD detectors without liquid N_2 cooling, a great improvement of portable XRF systems design has been achieved during the last decade.

2.4.4 DETECTOR ARTIFACTS

Additional peaks can arise in an XRF spectrum from sources other than the meaningful elements in the sample and it is important to be aware of their nature in order to get quality analytical results. These additional peaks are usually referred to as spectrum artifacts and some of them, namely escape peaks and sum peaks (pile-up effect), have their origin in the detector system.

2.4.4.1 Escape Peaks

The atoms in a detector (Ar, Si, or Ge) will also emit their own characteristic radiation when struck by the incoming X-ray beam. The majority of this radiation is immediately absorbed within the detector volume and contributes to the overall charge collected for the original incident X-ray photon. However, there is a limited probability that the X-rays produced will escape from the detector volume and not contribute to the charge collected for the original photon that was detected. The result of this escape process is a peak (escape peak) in the collected spectrum located at an energy value lower than the parent peak, equivalent to the difference between the energy of the original photon and the energy of the line of the detector element. As an example, the Si escape peak for Rh K_α is clearly seen at 1.74 keV below the Rh K_α peak, in Figure 2.15.

Usually, XRF instrument software makes a correction in which escape peaks are automatically removed. This is especially necessary to avoid classic errors in peak identification such as the Si escape peaks from Zn K_α (8.64–1.74 keV) being identified as Co K_α (6.93 keV).

2.4.4.2 Sum Peaks (Pile-Up Effect)

Sum peaks appearing in the XRF spectrum are the result of two events from high-intensity peaks that arrive in the pulse processing electronics of the detector so close together in time that

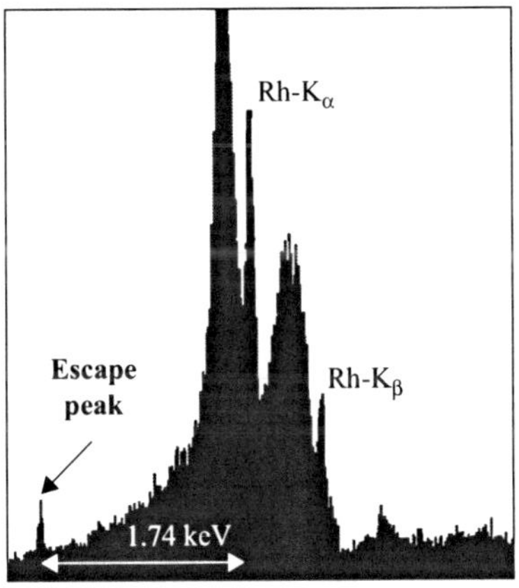

Figure 2.15. Escape peaks for a Rh X-ray tube radiation collected by a Si(Li) detector.

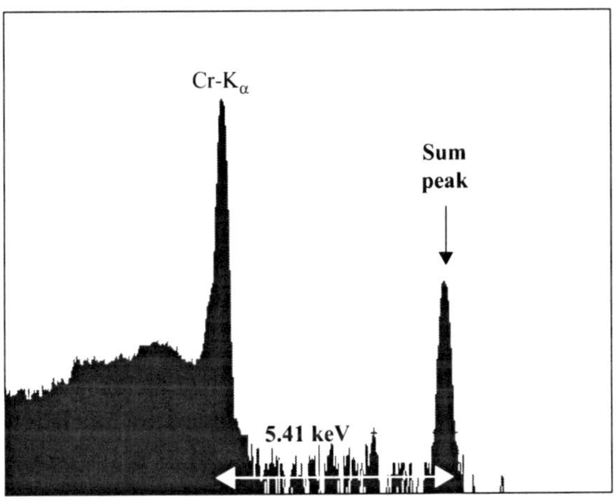

Figure 2.16. Graphical view of sum peak for Cr K line.

they are not recognized as two events but as only one. Therefore, the energy of the resulting peak is the sum of the two initial energies. This effect is inherent to the count rate limitations associated with the finite pulse processing time required by the electronic shaping network (pile-up effect). In Figure 2.16, the sum peak of Cr, which appears at two times the energy of Cr-K_α (2×5.41 keV) is shown. The sum of two Cr-K_β events is only visible when the parent peak is very intense.

In a similar way as that for dealing with escape peaks, XRF software packages are able to correct for sum peaks to minimize potential errors in element identification. This is particularly important in the analysis of some environmental samples with high Fe amounts (e.g., soil samples). This will result in Fe K sum peaks (2×6.40 keV) in the region of K lines of Se (11.2 keV) and Br (11.8 keV) and the L lines of toxic elements such as Pb (10.5 keV).

2.4.5 SIGNAL PROCESSING SYSTEM

The charged pulses produced by a detector are subsequently processed by a complex electronic system. They are amplified, shaped, and sorted according to their amplitudes. Finally, all pulses with certain amplitudes are counted. A strong proportionality between the amplitude of the

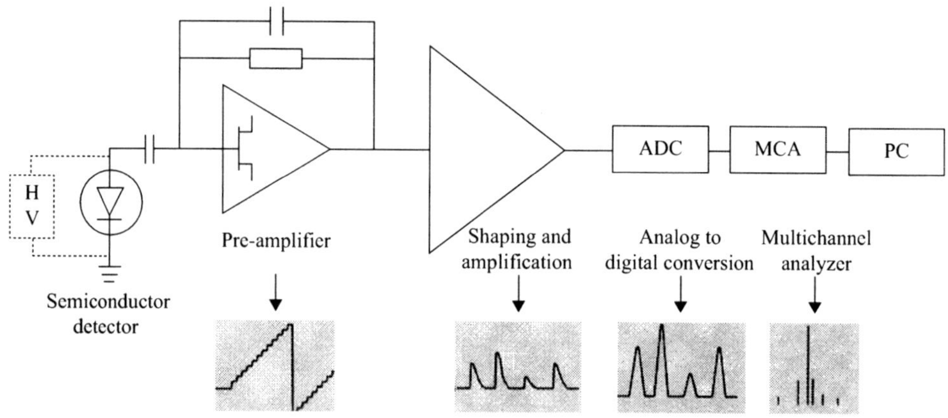

Figure 2.17. Scheme of signal processing in XRF instrumentation. HV, high voltage; ADC, analog-to-digital converter; MCA, multichannel analyzer; PC, computer.

initially produced pulses and subsequently processed pulses should be accurately maintained. As long as this condition is met, the pulse amplitude remains an accurate measure of the energy of the detected photons. Ultimately, the result of the photon counting can be represented as an energy dispersive spectrum representing the number of photons as a function of their energy. In Figure 2.17, a schematic of a signal processing system used in most XRF spectrometers is shown.

It is important to note that the recent development of digital pulse processors (DPPs) that digitize the preamplifier signal and then perform pulse shaping, provide lower electronic noise and better stability than an analog system at higher count rates. DPPs also integrate the shaping amplifier, MCA, and communication functions onto a single printed circuit board not much bigger than a credit card. Therefore, in combination with solid drift detectors, DPPs optimize the analytical performance of the XRF spectrometer in terms of both analytical range and sensitivity.

Spectrometer electronics need a certain amount of time to process a pulse during which no other pulse can be registered. This period is called *dead time* and is different for each detection system. As the measured pulses occur statistically distributed over time, the proportion of pulses occurring during the processing period of a previously registered pulse depends on the intensity of the radiation. This means that the greater the incident intensity, the higher the losses during the measurement dead time. Therefore, a software correction of the measured intensities is necessary to produce a linear relationship between incident and measured intensities.

2.5 SOURCE AND DETECTOR MODIFIERS

Most XRF instruments are equipped with one or more source and/or detector modifiers which are sited between the excitation source and the sample or between the sample and the detector, respectively, to improve the instrumental selectivity, sensitivity, or limits of detection for element determination.

Usually, the benefits of using source modifiers are related to a reduction of the continuum and the characteristic X-ray lines coming from the X-ray tube to avoid overlaps with the elements of interest or to increase the signal-to-noise ratio for element determination. In addition, some source modifiers are introduced to enhance the X-ray production by selecting appropriate excitation conditions or to decrease the size of the irradiating area to obtain local composition at different spots on the sample.

The detector modifiers are usually employed to collect the fluorescence radiation emerging from the sample and to modify its spectral distribution in favor of the lines of interest.

In the following sections, the design, operation, and benefits of the most commonly used source and detector modifiers are discussed.

2.5.1 FILTERS

Filters, usually thin metal layers (10–500 μm) that act as absorbers, can be placed between the X-ray source and the sample (primary filters) or between the sample and the detection system (detector filters).

2.5.1.1 Primary Filters

A primary beam filter is placed between the X-ray tube and the sample to modify the X-ray tube output spectrum to which the sample is exposed. It reduces the intensity of interfering lines and/or the continuum of the X-ray tube, improving in this way the signal-to-noise ratio for element determination. For instance, in Figure 2.18, an example is presented showing the benefits of using different primary filters to reduce the continuum radiation and the characteristic lines of a W anode X-ray tube operating at 50 kV for the analysis of a pure Si plate.

As can be seen in the case of Figure 2.18, for Zn and Pb determination, the use of a primary beam filter made of Al, Ti, or Mo is required to mostly eliminate the characteristic radiation of the W-L_α (8.4 keV) and W-L_β (9.7 keV) peaks from the X-ray tube which hampers the detection of these elements (Zn-K_α = 8.64 keV, Pb-L_α = 10.5 keV). For the determination of small quantities of heavier elements such as Cd (K_α = 23.2 keV), the best option to improve the signal-to-noise ratio is the use of a Ti or Mo primary filter. Another advantage of using primary filters is the absorption of part of the initial radiation to avoid the saturation of the detector.

The characteristics of a filter are defined by its X-ray absorption curve that is controlled by selecting the material and its thickness. In Table 2.4, some typical filters, the range of voltage used with each, and the optimum applicable element range are shown.

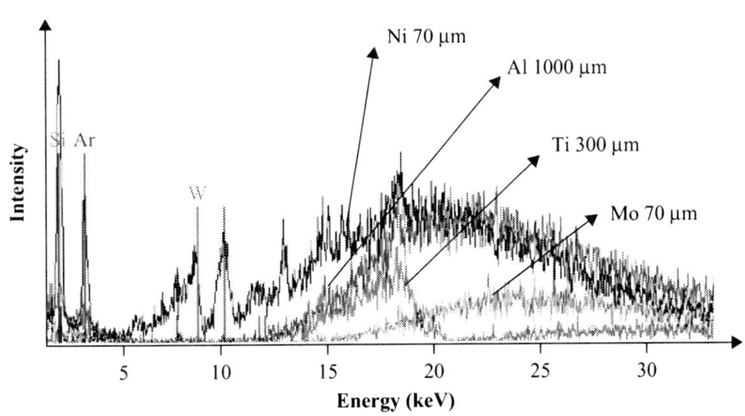

Figure 2.18. Influence of different primary filters on the X-ray tube spectrum output. W X-ray tube anode; sample: pure Si plate.

Table 2.4. Typical primary beam filters and their range of use (modified from Van Grieken and Markowicz (2002))

Filter	Thickness (μm)	Voltage range	Elements
Cellulose	single sheet	5–10	Si–Ti
Thin aluminum	27–75	8–12	S–V
Thick aluminum	75–200	10–20	Ca–Cu
Thin anode element	25–75	25–40	Ca–Mo
Thick anode element	100–150	40–50	Cu–Mo
Copper	200–500	50	>Fe

Most modern XRF systems provide the ability to use more than one set of conditions for analysis. Depending on the analytical requirements, the voltage and the filter are chosen to optimize the important parts of the spectrum and the data from the spectra at each of these conditions are consolidated to provide the best detection limit.

2.5.1.2 Detector Filters

Filters can also be introduced, for some specific applications, between the sample and the detector system to modify the spectral distribution of the fluorescence radiation emerging from the sample in favor of the lines of interest. For instance, the introduction of a Cr detector filter proved to be a good approach to improve the sensitivity for Se determination in soil digests characterized by high amounts of Fe. In view of the energy of the emission lines of Fe (K_α: 6.403 keV, K_β: 7.059 keV) and Se (K_α: 11.221 keV, K_β: 12.501 keV), a Cr foil with a thickness of 10 μm proved the best option to be used as an absorber as the absorption edge for this element (Cr-K_{abs}: 5.988 keV) is suitable for considerably decreasing the Fe peak (49.7%) intensity entering the detector without any significant reduction of Se signal (3.2%) (Marguí et al., 2010a).

2.5.2 SECONDARY TARGETS

Another option for modifying the X-ray tube output to which the sample is exposed is to introduce a secondary target between the X-ray tube and the sample. There are three types of targets of practical use: fluorescent targets, Barkla targets, and Bragg targets.

- **Fluorescence targets** are pure metal or compound targets that use their characteristic fluorescent radiation to excite the sample. These targets also scatter the tube radiation but this is lower than what is obtained with the direct excitation with an X-ray tube.
- **Barkla targets** are made of light-element compounds such as Al_2O_3 and B_4C. These targets are designed for efficient scattering and polarization of polychromatic source radiation from an X-ray tube.
- **Bragg targets** are crystals that reflect only one specific energy in a certain direction, thus reducing the background level in the recorded XRF spectrum.

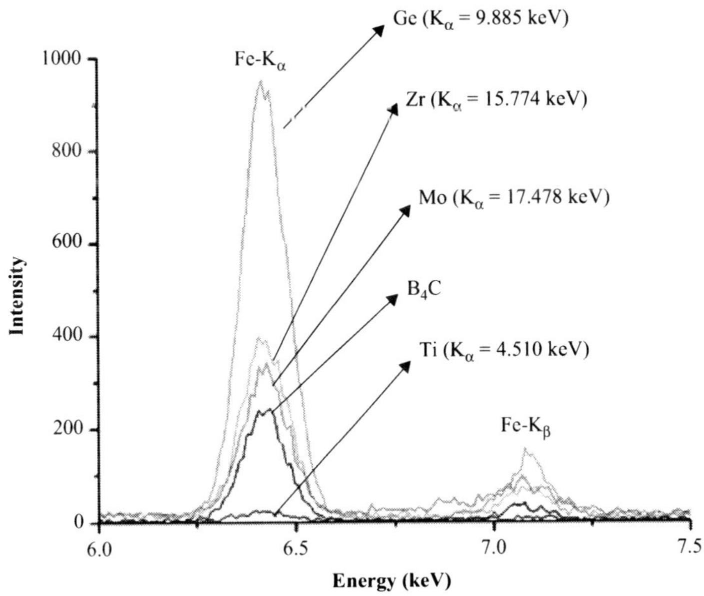

Figure 2.19. Influence of secondary target used for excitation on Fe determination.

The main advantage of scattering primary X-rays with targets is that it is much more practical to vary targets than X-ray tube anodes, and targets may be used to produce fluorescence lines to act as near-monochromatic secondary sources at slightly higher energies than the absorption edges of the analytical lines to be excited. By combining polarization with the possibility of different secondary sources, an instrument can be designed to generate a range of excitation and polarization conditions optimized for different groups of analytes. As an example, in Figure 2.19, we can see that in the specific case of Fe determination, we can obtain a better sensitivity when using a Ge secondary target.

2.5.3 FOCUSING OPTICS

Focusing optics devices are commonly used to focus the X-ray radiation onto a small spot of the sample (if they are introduced between the source and the sample) or to receive only its fluorescence from a small spot (if they are placed between the sample and the detector).

The simplest configuration to achieve a reduction of the irradiating or detecting area is the introduction of a pinhole. However, when using such systems, as is shown in Figure 2.20, one also gets a proportional reduction of the intensity of the primary X-ray beam. Therefore, its application is limited when elements with very low concentrations present in a sample have to be determined.

Another option is to use capillary lenses to shape and direct the primary X-ray beam. In these systems, the primary beam that impinges on the surface section of the lens propagates through the system and emerges in a convergent beam with high intensity (see Figure 2.20). At present, capillary optics present one of the fastest growing X-ray optical technologies because of their capacity to generate high-flux-density beams at the micrometer and submicrometer scale.

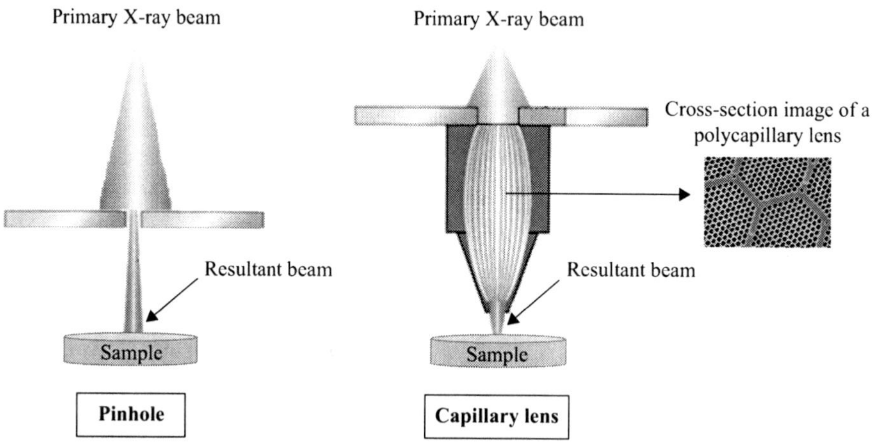

Figure 2.20. Resultant beams after placing a pinhole or a capillary lens as focusing optics between the primary X-ray beam and the sample.

The main idea of capillary optics is to use hollow channels in glass for transporting X-ray radiation. The simplest configuration is a monocapillary lens in which only a straight cylindrical bent hollow glass is used for focusing. More sophisticated and expensive configurations such as polycapillary systems using a group of fibers can also be used to obtain a larger intensity gain within the focal spot.

Different X-ray output diameters are required depending on the type of measurement. A combination of various optically prealigned capillaries with different output diameters (i.e., 10 and 100 μm) can be installed in an XRF system to facilitate changes between output diameters. Moreover, when using capillary lenses as focusing optics, one should be aware that the focal distance and the transmission of the lens is not the same for all energies (elements). For instance, very low energies are absorbed by the lens and very high energies are not focused but just pass through the lens. One has to take into account these effects depending on the elements of interest to be determined in the sample.

2.5.4 DISPERSING SYSTEMS

Dispersing systems (or monochromators) are usually crystals or synthetic multilayers (made of alternating layers of light and heavy elements or their compounds) that can be introduced between the X-ray source and the sample or between the sample and the detector to select a specific range of the X-ray spectrum (monochromatization). In the first case (primary radiation), they eliminate undesirable parts of a continuous spectrum or some of the characteristic lines, thus suppressing background and increasing correspondingly the sensitivity of the analytical method. In the second case, the dispersing system is applied for collecting the secondary (fluorescence) radiation emerging from the sample and modifying its spectral distribution in favor of the signal from the lines of interest entering to the detector.

The monochromatization of the radiation using these devices is done utilizing the Bragg reflection law. Fundamentals of this principle are shown in Figure 2.21. If parallel beams of X-rays fall on the crystal or multilayer, the first layer reflects a fraction of the X-rays. The remaining radiation penetrates the crystal and is reflected by the subsequent layers. If the

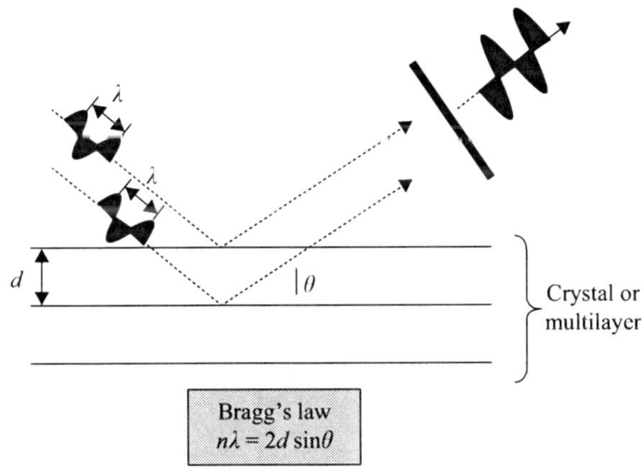

Figure 2.21. Bragg reflection law in a crystal or multilayer (dispersing systems) where n is an integer; d is the spacing between planes in an atomic lattice, or layer separation; λ is the wavelength; θ is the angle of incidence.

Table 2.5. Crystals and multilayers commonly used as dispersing systems

Crystal	Name, material	2*d* (nm)	Element range
LiF (220)	Lithium fluoride	0.284	V–U
LiF (200)	Lithium fluoride	0.403	K–U
Ge	Germanium	0.653	P, S, Cl
PET	Pentaerythrite	0.874	Al–Cl
TIAP	Thallium biphthalate	2.576	O–Mg
OVO-55	W/Si multilayer	5.5	Mg, Na, P

difference in the path lengths between reflections from layers is a multiple of half the wavelength of the radiation, the two reflection beams disappear (destructive interference). If the difference is exactly an integer times the wavelength, the two reflected beams reinforce (constructive interference). Therefore, depending on the orientation to the possible crystallographic planes of the crystal or multilayer, different values for d are preferentially diffracted and a different resolution results for a given wavelength of radiation.

In Table 2.5, a list is presented for some crystals and multilayers currently available and the element range of use.

2.5.5 COLLIMATORS

A collimator is a stack of parallel metal plates, spaced a few tenths of a millimeter apart, that are used to obtain a parallel beam of fluorescence radiation (see schematic in Figure 2.22). For instance, in combination with diffraction crystals, collimators are used to obtain a parallel X-ray beam that falls exactly at the required angle on the crystal. In such cases, a primary collimator is placed between the sample and the crystal, and a secondary collimator can be placed between the crystal and the detector.

In addition, to collimate the X-ray beam this arrangement has the advantage of simplicity and relatively low cost. However, it also reduces the intensity of the X-ray beam as part of the incoming radiation is lost. This phenomenon can be seen in Figure 2.23.

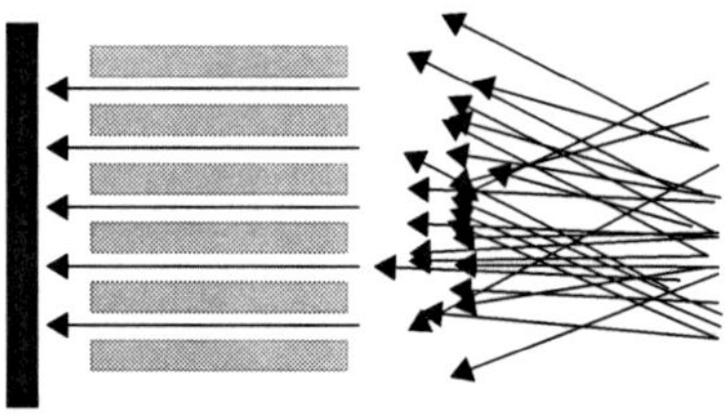

Figure 2.22. Schematic of a design and operation of a collimator device.

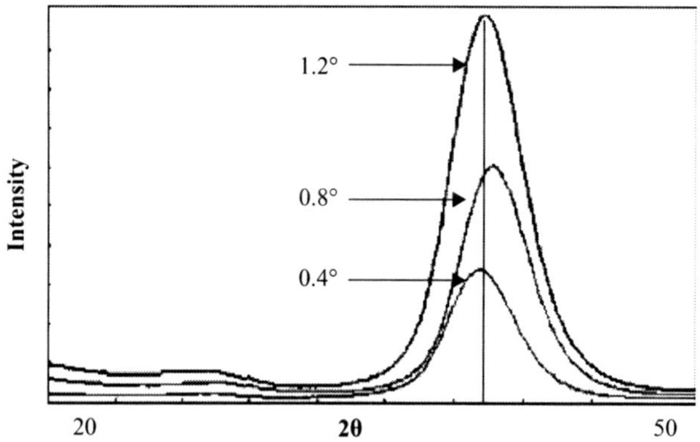

Figure 2.23. Influence of collimator aperture on the resulting radiation.

2.5.6 MASKS

Masks, which are plates with a hole in them, reduce the size of the fluorescence beam in front of the collimators or cut off X-rays emitted or scattered by an angle larger than the sample area. This latter case is especially useful for removing the radiation coming from the edge of the sample holder (cup) aperture. In such a case, the mask is placed between the sample and the detection system so that the detector only detects the radiation coming from the sample. The size of the mask is generally adapted to match the cup aperture being used.

2.6 INSTRUMENT CONFIGURATIONS

During the last decade, noticeable development was made in the instrumental aspects of X-ray spectrometry. This progress includes, for instance, considerable improvements in the design and production of detectors and X-ray optics. All this has resulted in a wide variety of instrumentation becoming available today—a long way since the first modern commercial X-ray spectrometer was introduced in the market at the end of the 1940s.

It is difficult to make a unique classification of XRF techniques, because, depending on the detection system mode, the excitation source, the geometry of the optical path, and also the spectrometer size, we can find several different instrument configurations. In Figure 2.24, a tentative classification taking into account these facts is proposed.

Usually, XRF spectrometers are divided into two main categories depending on the fundamentals of the detection system: wavelength-detection systems (WDXRF) and energy dispersive systems (EDXRF). WDXRF employs diffraction by a dispersing system (crystal

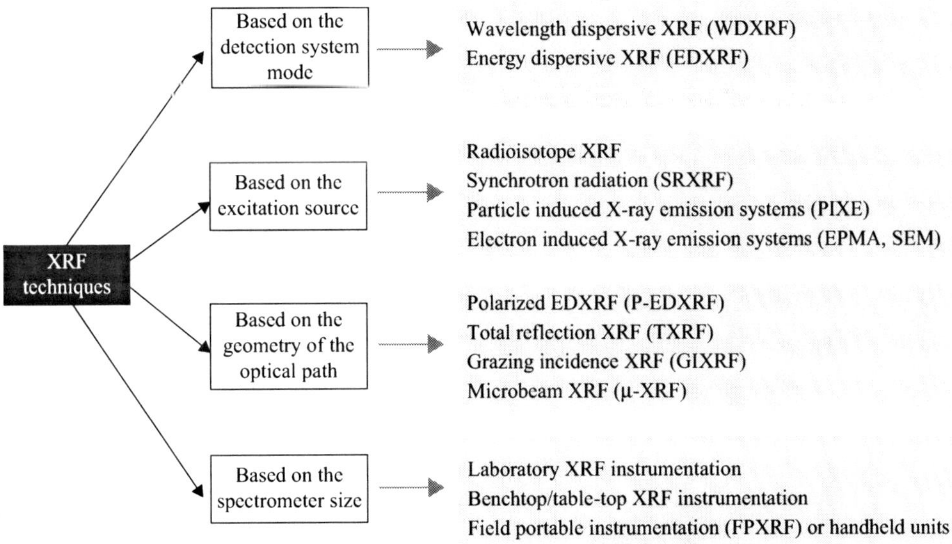

Figure 2.24. Classification of instrumental configurations in XRF techniques (and the related techniques: particle-induced X-ray emission (PIXE), electron probe microanalyzer (EPMA), and scanning electron microscopy (SEM)).

or multilayer) to separate the characteristic wavelengths emitted from the sample. In contrast, EDXRF spectrometers use the proportional characteristics of a photon detector to separate the characteristic photons emerging from the sample in terms of their energies. Taking into account that there is a simple relationship between wavelength and energy, these techniques provide the same basic type of information. The characteristics of the two configurations differ mostly in their relative sensitivities and the way in which data are collected and presented. In general, a WDXRF system is about one order of magnitude more sensitive than an EDXRF instrumentation. However, an EDXRF spectrometer measures all elements within its range at the same time, whereas a WDXRF system identifies only elements for which it is programmed (using different dispersing devices). In Chapter 5, we will discuss in more detail the basics, applications, and differences between WDXRF and EDXRF systems.

As mentioned in Section 2.2, the most common radiation sources used in X-ray spectrometry are X-ray tubes. Nevertheless, other sources can be used to irradiate the sample including radioisotope sources, synchrotron radiation, and particle beams. The main principles, benefits, and particular fields of application of these sources will be summarized in Chapter 7 that will also focus on special XRF configurations and related techniques. In Figure 2.25, an overview of the incidence and emerging beams is presented.

Depending on the setup (geometry) of the basic components of the spectrometer, one has also differently configured XRF systems. For instance, in polarized-EDXRF instrumentation (P-EDXRF), the X-ray optical path between the X-ray source, the sample, and the detector is not limited to one plane, like in conventional EDXRF systems, but involves three-dimensional optics (3D optics). The main advantage of this geometry is that scattered tube radiation cannot reach the detector because of polarization and thus the sensitivity is improved compared to 2D optics. On the other hand, in total reflection XRF (TXRF) and grazing incidence XRF (GIXRF) systems, the primary beam strikes the sample at a very small angle ($\leq 0.1^\circ$) compared to a

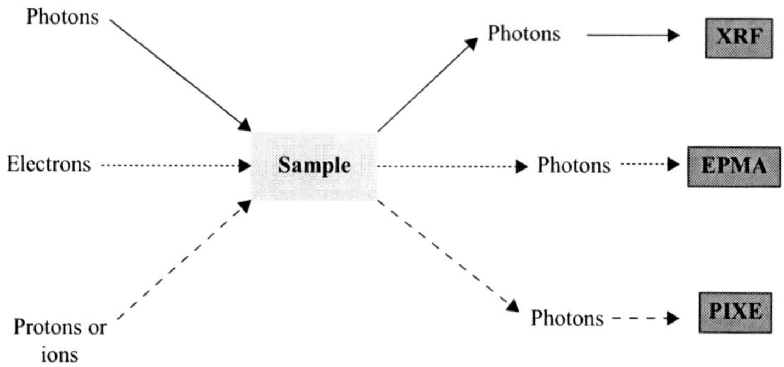

Figure 2.25. XRF-related techniques depending on the incident sources.

conventional EDXRF systems (~45°) and that results in a low penetration depth. Consequently, both techniques have interesting applications in the analysis of thin layers. In Chapter 6, a detailed discussion on the TXRF configuration and related applications is presented. Finally, in some XRF spectrometers, a focusing optic device is introduced between the X-ray source and the sample to focus the incident radiation onto a small spot on the sample in the micrometer and submicrometer size scale (μ-XRF). Originally, micro X-ray analysis systems were largely restricted to charged-particle beam techniques, but significant advances in focusing optics (mostly the development of polycapillary lenses) have promoted the development of microbeam XRF (μ-XRF) instrumentation employing (conventional) X-ray tubes as the source of primary radiation. Nowadays, μ-XRF is an indispensable tool in many disciplines (including industrial, environmental, cultural, and forensic) to get information on the element distribution at the micrometer and submicrometer scale. Additional information on this type of XRF instruments will be discussed in Chapter 7.

Over the last few decades, most of the published XRF analyses were performed using large-scale instruments with high-power X-ray tubes, which demanded water-cooling systems and liquid-nitrogen cooled detectors. However, in recent years, the development of ceramic low-power X-ray tubes (maximum power ~50 W) and SDDs, that do not require cooling media and gas consumption to operate, has promoted the development of benchtop or table-top XRF instrumentation (both EDXRF and WDXRF) in low-cost compact designs. These improvements in X-ray excitation and detector systems, coupled with technological development in batteries and microprocessors, have led to the adoption also of field-portable (FPXRF) or handheld units for many analytical purposes. These devices allow almost point-and-shoot ease of use and can offer results comparable to those of laboratory-based instruments.

CHAPTER 3

Qualitative and Quantitative X-Ray Fluorescence Analysis

3.1 EVALUATION OF X-RAY FLUORESCENCE SPECTRA

After the acquisition of an XRF spectra, the spectrum itself is just a data set and not yet a result. It is an iterative process of spectra correction and evaluation that leads to providing the solution to the analytical task from a spectrum, which can be focused on obtaining information on the elements present in the target sample (qualitative analysis) and/or on the quantification of the amount of the elements present in the target sample (quantitative analysis). In Figure 3.1, a brief summary of the different steps involved in spectra evaluation is shown.

- **Spectrum correction**
  As already stated in Chapter 2, in addition to the characteristic lines of the elements present in the sample, other peaks can arise in an XRF spectrum. It is important to be aware of their origin and their location within a spectrum to get meaningful, quality analytical results. In this category we have to include:
  - Continuum radiation and Rayleigh and Compton scatter peaks arising from the interaction between the primary radiation and the sample (see Section 1.2)
  - Escape peaks and sum peaks arising from the detector system (see Section 2.4.4)
  - Other artifacts include characteristic lines from elements added to the sample as a consequence of the sample treatment procedure: those arising from the measurement atmosphere (i.e., Ar from the air if we do not work under vacuum), those coming from the sample holder, and so forth.

Usually, XRF instrument software makes a correction by which the escape and sum peaks are automatically removed. However, we have to consider the identification of other artifact peaks to make unequivocal the identification of the elements present in the analyzed sample.

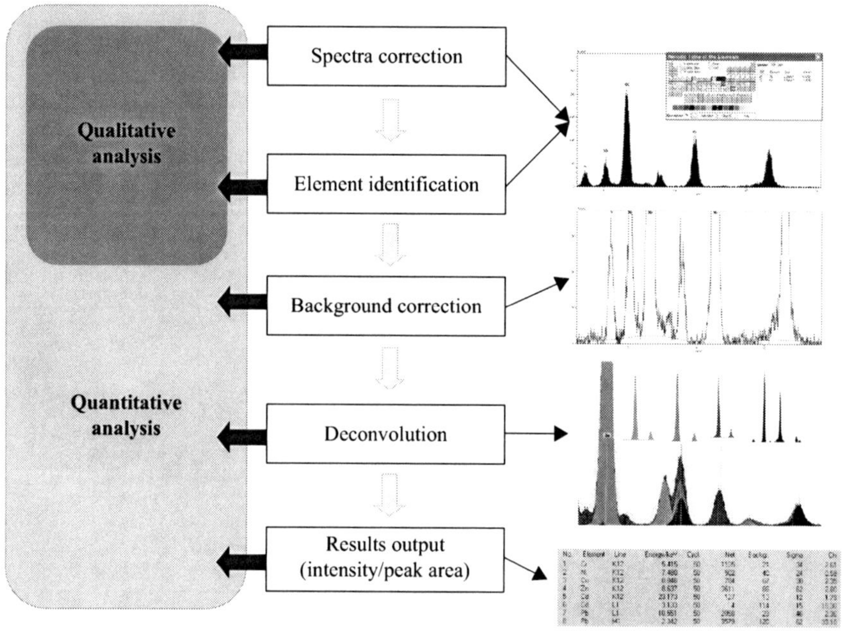

Figure 3.1. Processes involved in the evaluation of XRF spectra.

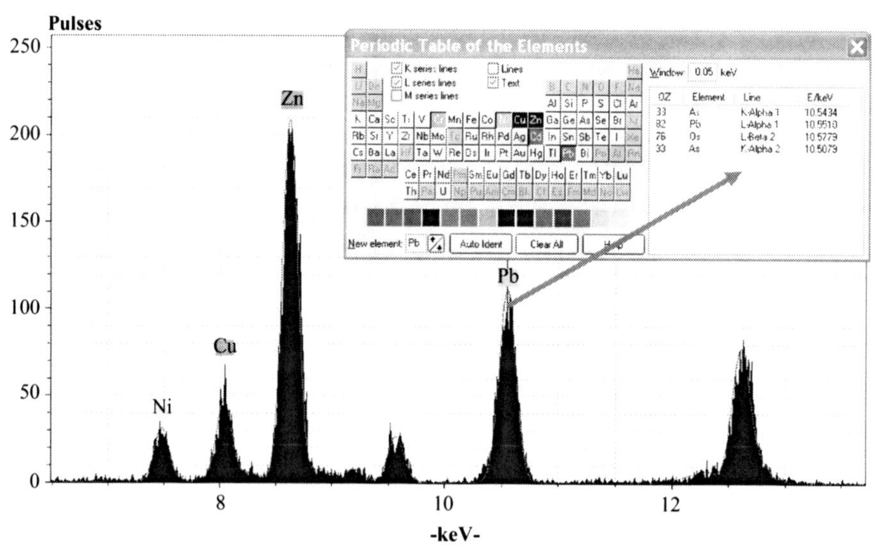

Figure 3.2. Automatic element identification processes, for example, identification of a Pb peak. (Atomic data library S2 PICOFOX, Bruker AXS).

- **Element identification (Qualitative analysis)**
  The identification of the element peaks present in a spectrum is usually performed by using the software of the XRF instrument. The operation mode consists of a comparison between the spectral lines appearing in the spectrum and the lines of the individual elements stored in the software in the form of an atomic data library. In Figure 3.2, an example of an automatic element identification is shown.

Therefore, by element identification we can get a first qualitative analysis of the atomic constituents of the target sample. If a quantification is needed, then element peaks must first be corrected for the spectral background and possible overlaps.

- **Background correction**
  In both WDXRF and EDXRF, the concentration of an analyte is proportional to the number of counts under its characteristic X-ray peak (analytical signal); therefore, any background signal must first be subtracted from the spectrum. Usually, in EDXRF, the analytical signal used is the peak area, while in WDXRF systems it is the peak height.
  The background, which is composed of detector noise and scattered excitation radiation, is calculated by means of a mathematical algorithm (often some polynomial type of function) and subtracted from the spectrum as can be seen, for example, in Figure 3.3.
- **Deconvolution**
  For multielement analysis in particular, strong overlaps of the individual fluorescence peaks in a spectrum occur because of the finite resolution of the detection system, especially in EDXRF for which the resolution power is dependent only on the detector. For this reason, mostly in EDXRF, to extract the correct intensities of the individual elements, a spectrum must be separated into the individual element lines by a complex mathematical process called *deconvolution*. The result is the net analytical signal for each element of the individual, *nonoverlapped* fluorescence peaks.
  There are different procedures to allocate the final analytical signal for each element. In some softwares, the deconvolution step uses measured monoelement profiles for the evaluation of peak intensities. Alternatively, the fluorescence peaks can also be obtained by a mathematical Gaussian function.
  Usually, to get information on the quality of the deconvolution procedure, a statistical parameter called *Fit quality (Chi)* is calculated taking into account the standardized square sum of the differences between the measured and the deconvoluted intensities.

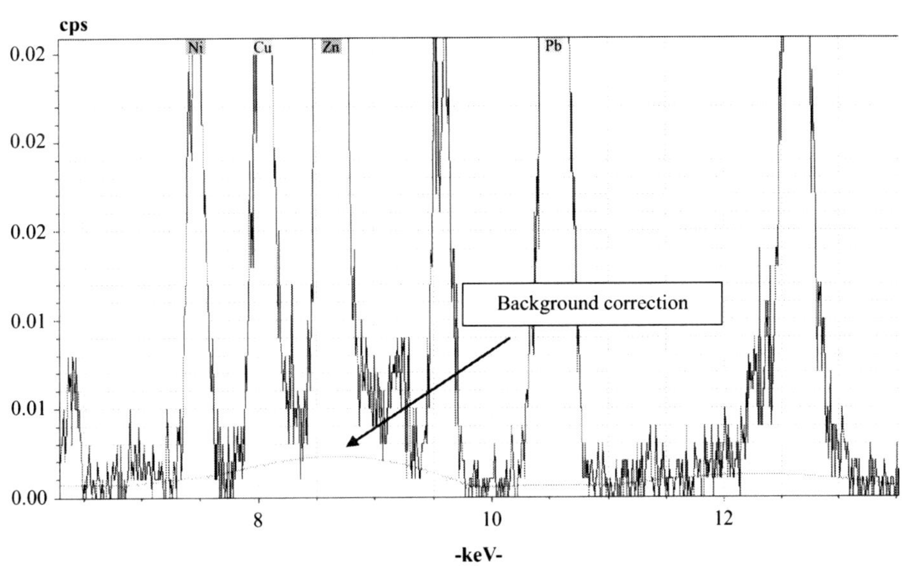

Figure 3.3. Background calculation of an EDXRF spectrum. (S2 PICOFOX, Bruker AXS).

The value for the fit quality should preferably be smaller than 10 (it depends on the software recommendations). Higher values are an indication of a poor fitting of the XRF spectrum. An example of that is displayed in Figure 3.4.

- **Results output**

  After a spectrum deconvolution, analytical results are usually displayed as net peak areas (EDXRF) or net peak height values (WDXRF). The obtained net analytical values are used then to estimate the relative concentration of the elements present in the sample (see Section 3.3).

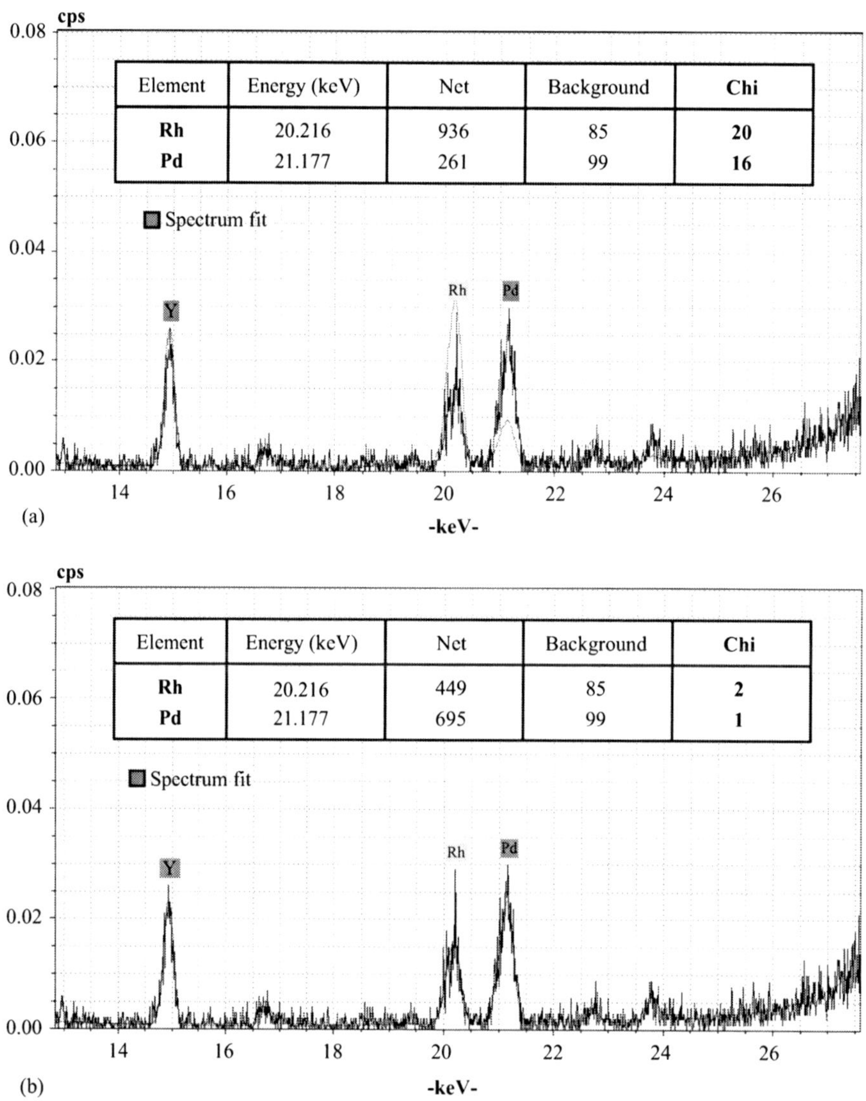

Figure 3.4. Deconvolution procedure and associated fit quality value. (a) Poor fit, (b) good fit.

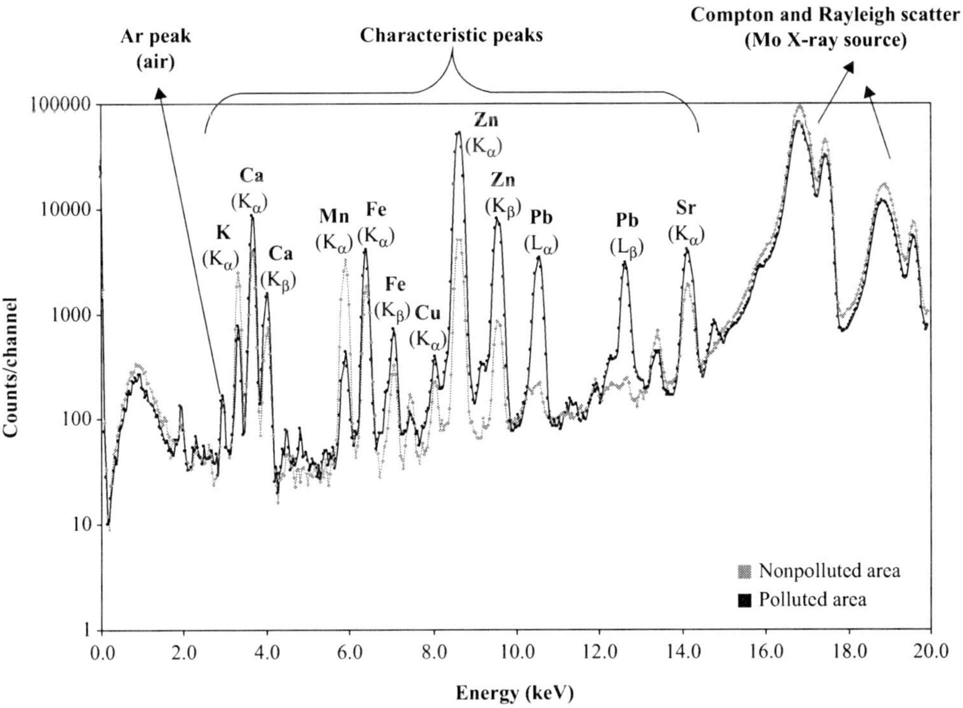

Figure 3.5. Comparison of EDXRF spectra for vegetation leaves collected from a metal-polluted area (mining area) and a nonpolluted area (modified from Marguí et al. (2005a)).

3.2 QUALITATIVE XRF ANALYSIS

As already stated above, qualitative information can be derived directly from the recorded XRF spectrum. However, one must still also identify other peaks appearing in the spectrum that are not coming from the sample (i.e., scatter peaks, sum peaks, escape peaks, peaks coming from the measurement atmosphere, etc.) in order to get meaningful analytical results. Figure 3.5 shows an example of qualitative analysis of vegetation leaves collected from a metal-polluted area (Pb/Zn mining area) and a nonpolluted area (control sample). As can be seen, at a cursory glance, the comparison of the spectra readily allows discrimination between the polluted and nonpolluted specimens. This kind of information is valuable in environmental studies to get a first approximation of the composition of the studied samples.

3.3 QUANTITATIVE XRF ANALYSIS

Quantitative XRF analysis involves the conversion of the net analytical signal (corrected for background and line overlap, see Section 3.1) to the concentration of analytes by a calibration curve method, obtained earlier from analyzing many standard samples. Ideally, the intensity of an analytical line is directly proportional to the concentration of the analyte in the sample. However, this relationship is not generally linear, as it depends on both physical and chemical matrix effects. The former occurs as a result of variations in the physical characteristics of the sample, including particle size, uniformity, homogeneity, and surface condition. To minimize these effects, a thorough and consistent sample preparation is crucial, as will be discussed later in Chapter 4.

3.3.1 CHEMICAL MATRIX EFFECTS

Chemical matrix effects result from differences in concentrations of interfering elements present in the sample. These interelement effects fall into two broad categories: absorption effects and enhancement effects.

3.3.1.1 Absorption Effects

Both, the excitation radiation and the X-ray emission are attenuated (absorbed) while traveling through the sample. Primary absorption occurs because all atoms of the specimen matrix will absorb or scatter photons from the primary source. Therefore, the portion of the spectrum that is most effective in the excitation of a given analyte is modified depending on the other matrix elements present. Secondary absorption refers to the effect of the absorption of the characteristic analyte radiation by the specimen matrix. As characteristic radiation passes out from the specimen, from which it was generated, it will be absorbed by all matrix elements to extents that relate to the mass absorption coefficients of these elements (see Figure 3.6).

In addition to the matrix composition, the magnitude of the absorption depends on:

- energy of the radiation;
- path length *d* (see Figure 3.6) of the atoms that have to be passed;
- density of the sample.

The absorption increases as the path length, density, and atomic number of the elements in the sample increase, and as the energy of the radiation decreases. The absorption can be so high that elements deep within the sample are not reached by the incoming radiation, or the characteristic radiation can no longer leave the sample and arrive at the detector. Table 3.1 gives the approximate analysis depth in various materials for three element lines with different energies.

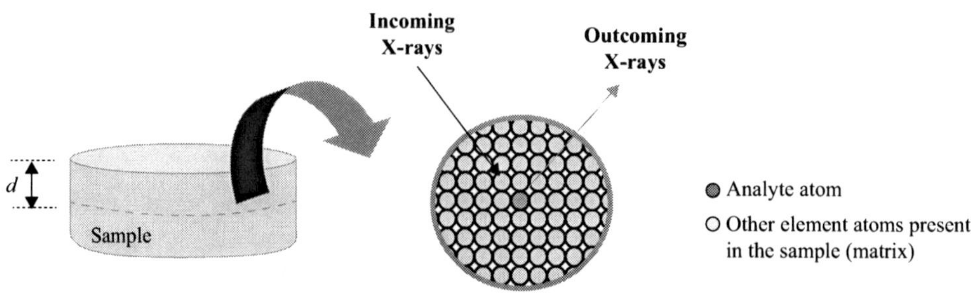

Figure 3.6. Absorption of incoming and outcoming fluorescence X-ray beams. (Note that the alignment of element atoms is not as regular as depicted in the diagram).

Table 3.1. Analysis depth for three different lines and various materials

Line	Energy (keV)	Material		
		Graphite	Iron	Lead
Mg-K_α	1.25	20 μm	1.92 μm	1.13 μm
Ni-K_α	7.48	4.39 mm	0.31 mm	16.6 μm
Cd-K_α	23.17	14.5 cm	0.70 mm	77.3 μm

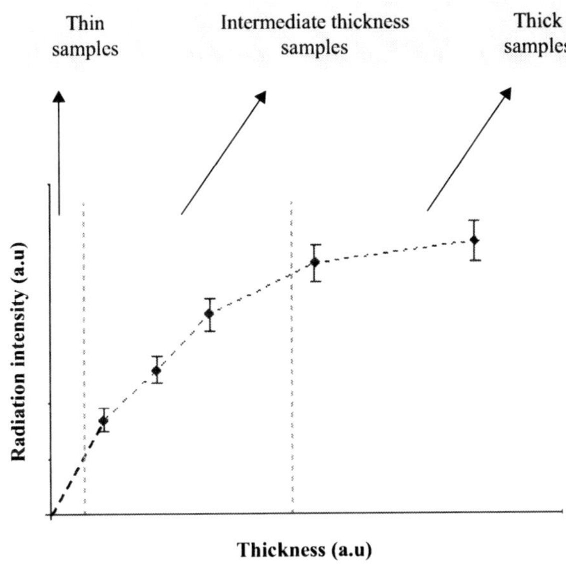

Figure 3.7. Relationship between radiation intensity and sample thickness. a.u., arbitrary units.

As is shown, the analytical information for the determination of Mg in a Pb matrix comes only from the first micrometer near the sample surface (~1.13 μm). However, if we are interested in Cd determination in a light matrix (graphite), we can obtain analytical information from a depth up to 14.5 cm from the surface.

As a consequence, there is a relationship between radiation intensity and sample thickness. From a practical point of view, depending on the effective thickness of the measured sample, one can work under regimes of "thin samples," "intermediate thickness samples," and "thick samples" (see Figure 3.7).

- **Thin samples**

When the mass per unit area is very small (thin sample), it is possible to consider that the total mass absorption coefficient of the sample is almost negligible (see Figure 3.8) and a linear relationship between the intensity of the fluorescent radiation and the concentration is observed.

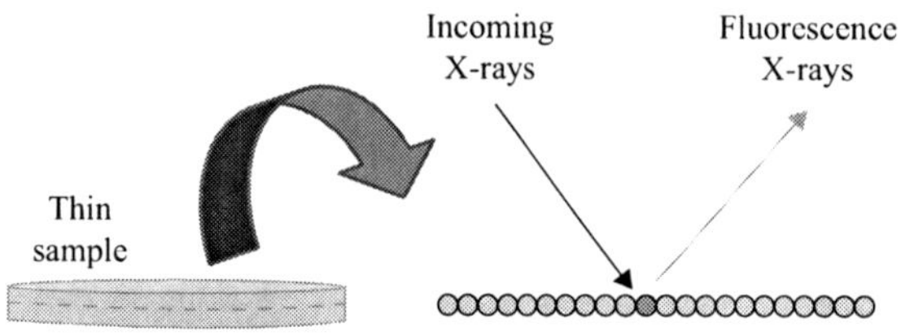

Figure 3.8. Absorption of incoming and fluorescence X-rays is negligible in thin samples.

- **Intermediate thickness samples**

Here, the potential thickness from which an element can be measured is greater than the thickness of the sample and the relationship between the measured fluorescence and element concentration is no longer constant. When working with this type of sample, a particularly useful technique used to avoid errors in quantitative measurements is to weigh a constant mass of sample on which to perform the measurement (to attain a constant sample thickness).

- **Thick samples**

At a critical thickness, secondary X-rays are excited at depths from which they cannot reach the surface and thus, a further increase in thickness results in no further increase in characteristic X-ray intensity (effect of sample thickness becomes negligible).

Therefore, taking into account the significant effect of sample thickness on the analytical signal, whenever possible, the best option for quantitative analysis is to prepare the sample as a thin layer. Quantitative results can then easily be obtained directly from the peak intensities. However, in most cases, this is not possible and samples must be prepared as thick samples or intermediate thickness samples. We will discuss in detail the sample treatment procedures in Chapter 4.

3.3.1.2 Enhancement Effects

Enhancement effects occur when a nonanalyte matrix emits a characteristic line that has energy just in excess of the absorption edge of that of the analyte element. This means that the nonanalyte element in question is able to excite the analyte, generating characteristic photons over and above those produced by the primary continuum. This gives an increased, or enhanced, signal from the analyte. A graphical view of this effect is shown in Figure 3.9.

A spectrometer will measure the sum of both the primary and secondary fluorescences, and it is impossible to distinguish between the two contributions. In such cases, correction methods have to be used to obtain quantitative analytical results.

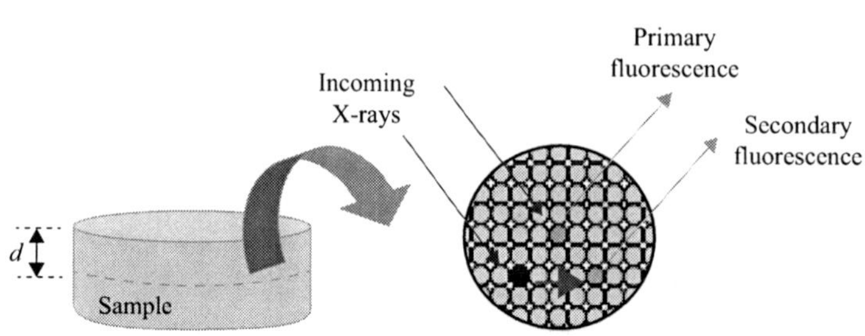

Figure 3.9. Schematic of the enhancement effect.

3.3.2 CORRECTION AND COMPENSATION METHODS

Ideally, the intensity of an analytical line is directly proportional to the concentration of the analyte in the sample. The proportionality constants (Q_i and q_i, see Figure 3.11) consist of many physical and instrumental factors, such as the probability of an atom emitting its characteristic radiation and the probability of these photons being detected and registered. However, the relationship between the intensity of an analytical line and the concentration of the analyte is not linear in the more general case owing to the matrix effects mentioned earlier. In Figure 3.10, examples of deviations from linearity due to variations in matrix effects are illustrated.

Therefore, the contribution of matrix effects must be taken into account in the expression of the radiation intensity in XRF to obtain meaningful quantitative results (see Figure 3.11).

The absorption effect is defined by the total mass absorption coefficient, the density, and the thickness of the sample. With regard to the enhancement effect, it is represented by means of an enhancement term, see Figure 3.11, that will be described in further detail later in this section.

Depending on the type of sample being analyzed, the expression of the radiation intensity can be simplified. For instance, as stated in the previous section, in thin samples it is possible to consider that the total mass absorption coefficient is almost negligible and thus the contribution of the absorption effect can be removed from the expression. However, when analyzing thick and intermediate thickness samples, correction models have to be used to obtain meaningful quantitative analytical results.

Several methods have been described for matrix effect correction. The most commonly used are displayed in Figure 3.12.

Each of these methods has its own advantages and disadvantages and the optimum choice is determined by each particular application. Usually, if the objective of the quantitative analysis is the determination of a single or several elements in an unknown but relatively constant

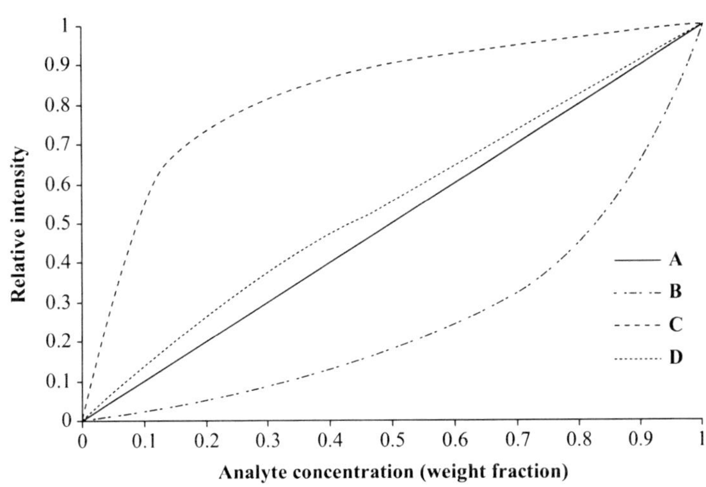

Figure 3.10. Effect of matrix effects (A) negligible matrix effect, (B) net absorption of the analyte's radiation by the matrix (positive absorption), (C) net absorption of the analyte's radiation by the analyte (negative absorption), (D) enhancement of the analyte's radiation by the matrix (modified from Van Grieken and Markowicz (2002)).

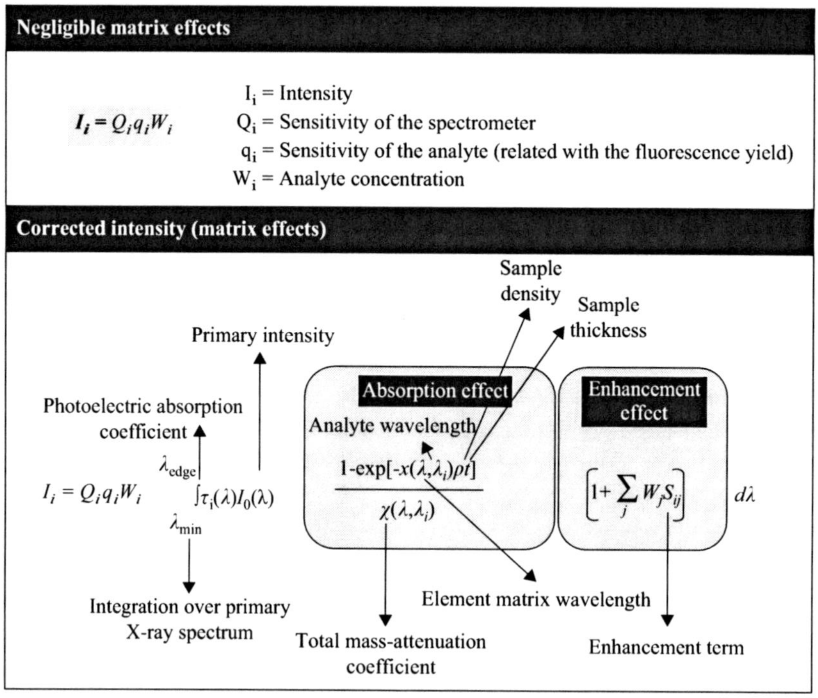

Figure 3.11. General relationship between intensity and concentration in XRF.

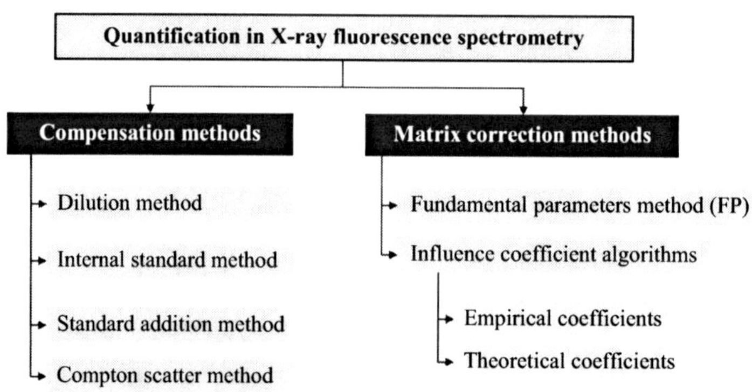

Figure 3.12. Compensation and matrix correction methods used in quantitative XRF analysis.

matrix, compensation methods are preferred. On the other hand, if the matrix effect is more variable from one sample to another and/or multielement determination is required, the use of the so-called *mathematical methods* (correction methods) is mandatory. Matrix correction models use terms to correct for the absorption or enhancement effects of the other elements and they are divided in two main categories: fundamental parameter (FP) method or influence coefficient algorithms.

3.3.2.1 Compensation Methods

Four different compensation methods may be used for quantification purposes in XRF analysis. The selection of the best one to use is dependent on the type of sample, analyte type, and concentration.

- **Dilution method**

Dilution methods can eliminate or reduce the variation of the matrix effect, rather than compensating for such variations. As dilution requires the addition of reagents to the sample, its use with powdered samples is limited owing to inherent problems of obtaining homogeneity. In addition, the application of this methodology to trace element determination is detrimental because with the dilution procedure the analyte intensity will also be reduced.

- **Internal addition method**

In this method, an element is added to each sample in a fixed proportion to both standards and unknowns to minimize the effect of variations in matrix effects. The element added (internal standard) should be similar to the analyte in terms of its absorption and enhancement properties. As the internal standard is affected in much the same way as the intensity of the analyte, provided there are no absorption edges or characteristic tube lines between the two lines considered, it can be used to correct for both absorption and enhancement. In such a case, a linear relationship between the intensity of the analytical line and the concentration of the analyte can be found using the following equation:

$$\frac{I_i}{I_s} = \frac{K_i}{K_s}\frac{M_i}{M_s}\frac{W_i}{W_s} \qquad (3.1)$$

$$\boxed{\frac{I_i}{I_s} = K_{is}W_i} \quad K_{is} = \frac{K_s M_s}{K_i M_i W_s} \qquad (3.2)$$

Where I_i and I_s are the analytical signal of the analyte and the internal standard, K_i and K_s are the calibration constants for the analyte and the internal standard, M_i and M_s are the matrix effects for the analyte and the internal standard, and W_i and W_s are the analyte and internal standard concentrations, respectively. Taking into account that the same amount of internal standard is added to each standard and sample and that M_i/M_s is a constant value, we can directly relate the ratio of the intensity of the analyte and the internal standard with the concentration of the analyte (expression 3.2).

The internal standard method is quite simple but it has also several drawbacks:

- The internal standard has to be added to each of the standards and samples. Therefore, sample preparation is more complicated and is more subject to errors.
- The addition of reagents and the requirement of homogeneity of the sample limit the practical application of the method to the analysis of liquids and fused specimens.
- In samples where many elements are to be quantified, several internal standards have to be used to provide a suitable internal standard correction.

- **Standard addition method**

Sometimes an appropriate internal standard cannot be found and, in this instance, it may be possible to use the analyte itself as an internal standard. The technique of standard addition is especially useful for the determination of a unique analyte present at concentrations below 5 percent. The fundamentals of the method consist of adding a known amount of the analyte (ΔW_i) to the sample and then the intensity of the analyte (I_i) will be increased by an amount ΔI_i. If we assume a linear calibration relationship, we can express the concentration of the analyte (W_i) as:

$$W_i = K_i I_i \tag{3.3}$$

$$W_i + \Delta W_i = K_i \left(I_i + \Delta I_i \right) \tag{3.4}$$

However, care must be taken in the application of the method whenever the relationship between analyte concentration and analyte intensity is nonlinear. To ensure linearity, it is a common practice to repeat the standard addition step at least twice so that a minimum of three data sets (zero concentration plus two different concentration additions) are available to work with. The concentration of the analyte in the sample is found by extrapolating the "calibration" line to 0. In Figure 3.13, an example of the standard addition method applied to selenium determination in soil digests by TXRF is shown (Marguí et al., 2010a).

Similarly to the internal standard method, the standard addition method also has several negative aspects. The most important one practically is the requirement to perform a calibration (using a minimum of two additions) for each sample.

- **Scattered radiation method**

As stated in Chapter 1, in addition to the already described photoelectric absorption, X-rays can also undergo scattering through elastic interaction (coherent or Rayleigh scatter) or inelastic interaction (incoherent or Compton scatter) with the outer electrons in the atoms. The intensity of the scatter is dependent on the energy of the incident ray and the average atomic number of

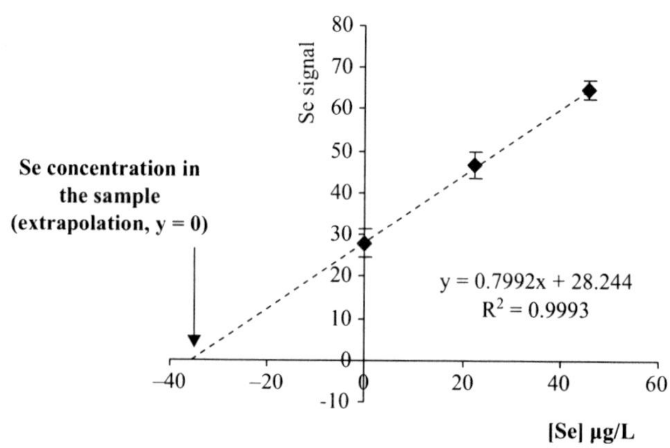

Figure 3.13. Example of standard addition method for Se determination in a soil extract (modified from Marguí et al. (2010a)).

elements of the matrix. Therefore, if the variation in matrix effects is mainly due to absorption, scattered X-rays can be used to correct the intensity of the analyte. The method also corrects, to some extent, for grain size and surface effects (disturbances of the fluorescent intensity due to the physical characteristics of the prepared specimens) and variations in tube voltage and current. However, it is not useful to correct for enhancement effects and thus it is limited in application to analysis of analytes that are influenced by absorption only. Another limitation of the method is that there should be no absorption edges of major elements between the analytical wavelength and the wavelength of the scattred radiation (to ensure that the ratio between the mass attenuation coefficients for those wavelengths remains constant between specimens).

In practice, the scattered radiation method is quite simple and consists of plotting the corrected intensity of the analyte ($I_i/I_{\text{scattered radiation}}$) versus the concentration of the analyte to obtain a calibration. This can be used for both coherently and incoherently scattered primary radiation as well as the scattered continuous radiation.

As scatter increases with decreasing mean atomic number, it is particularly suitable for matrix correction of elements within light matrices. The most common application of this approach is to determine the concentration of trace elements in specimens of geological origin; an example is presented in Figure 3.14.

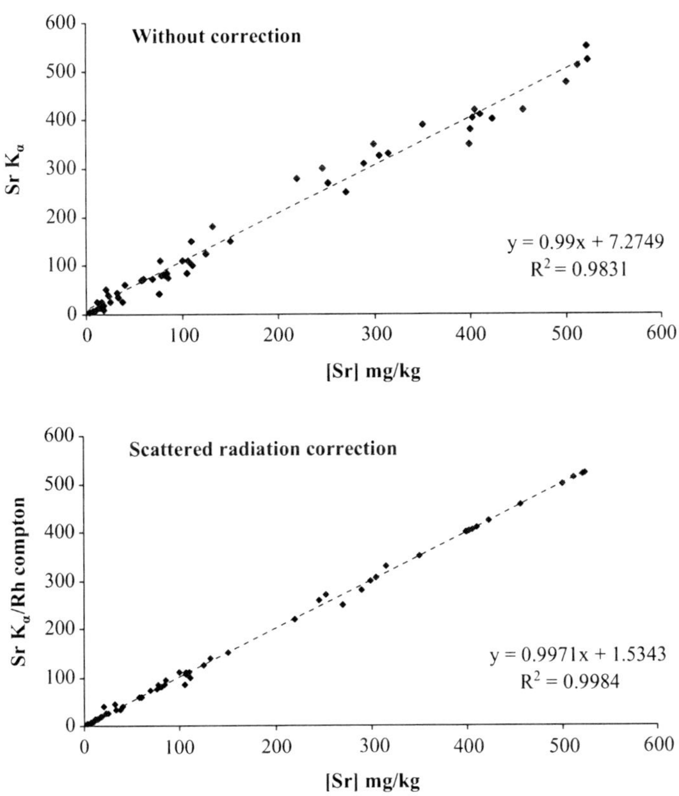

Figure 3.14. Example of the benefit of using scattered radiation correction method; top, without correction; bottom, with correction. (Sr determination in samples of geological origin) (modified from Van Grieken and markowicz (2002)).

3.3.2.2 Matrix Correction Methods

Matrix correction methods are based on mathematical models that calculate rather than eliminate matrix effects. When these correction methods are used, prior knowledge of the sample matrix is required so that one can get information on all measurable elements in the sample (multielemental information).

The mathematical methods are divided in two main categories: the fundamental parameters method and the methods using influence coefficients (Sitko & Zawisza 2012).

- **Fundamental parameters (FP) method**

The fundamental parameters method is based on the physics of X-rays and consists of obtaining the theoretical composition of the sample from the measured intensities. The mathematical equations used consider both primary (absorption) and secondary fluorescence (enhancement) and also take into consideration many physical constants and instrumental parameters (see Table 3.2).

In Figure 3.15, a summary of the steps that should be taken into account to calculate the sample composition following the FP methodology is shown.

In the first step, the measured intensity is converted to a theoretical intensity scale by using the instrumental sensitivity, and thereby an initial estimation of the composition is determined. Next, the theoretical intensity is calculated on the basis of the estimated composition and compared with the measured intensity. The concentrations are subsequently adapted: A convergence check for the concentration is made, and if the difference between the previous and the current concentration is reduced to a certain value, the resultant concentration is considered to be the final one. If not, the operation returns to calculation of the theoretical intensity until the required convergence has been attained.

Table 3.2. Physical constants and instrumental parameters considered in FP method

Physical constants
Wavelengths of absorption edges and emission lines
Fluorescence yields
Ratios of intensity of different lines within a given series
Mass-attenuation coefficients
Photoelectric absorption coefficients
Instrumental parameters (incident beam and geometry)
X-ray tube: target, voltage, current, Be window thickness
Primary filter
Secondary target
Incidence and exit angles
Distance: X-ray tube/sample/detector

It is interesting to note that before the application of these steps, an instrumental calibration procedure is necessary to calculate the sensitivity factors needed to convert the measured intensity to the theoretical intensity (step 1 (calibration) in Figure 3.15). The instrumental sensitivity can be determined by means of the analysis of pure element standards. This approach is especially useful if insufficient standards are available for calibration purposes. However, the use of one or a set of standards similar to the unknown samples almost always leads to more accurate results. This is because most of the uncertainties, caused by inaccuracies in the physical constants (10%–20%), are reduced. In a general way, the relative precision and accuracy of the results using a set of similar standards are usually better than 5%–10%, which is sufficient for most analytical applications.

Some disadvantages of the FP method are that it does not take into account some important physical processes in the sample such as scatter of both the primary radiation and fluorescence radiation, and also the photoelectrons and Auger electrons that are produced as a result of absorption of the primary and fluorescent X-ray photons (this contribution is significant in the case of low-Z elements).

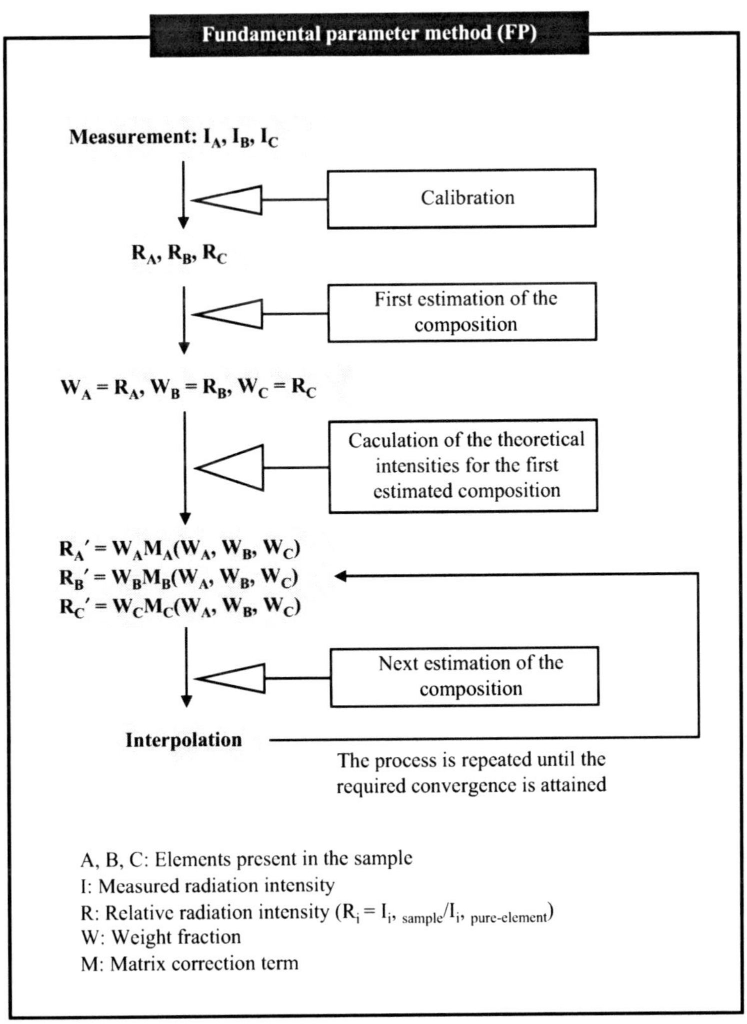

Figure 3.15. Fundamental parameters (FP) steps.

In addition, to obtain quantitative results using the FP approach, all elements present in the sample have to be detected. This is difficult when light-matrix samples have to be analyzed (owing to the low fluorescent yield of low-Z elements). In such cases, a variant of the FP method called *Backscattered fundamental parameter method* (BFP) is commonly used. In this model, the corrections for matrix effects are calculated using a combination of the estimated concentrations of all detected elements with a hypothetical "dark matrix," composed by the elements not detected by XRF. In practice, an experimental calibration of the ratio of incoherent to coherent scatter peaks for measured compounds (with known average atomic number) serves to establish an equation to determine the sample effective number for unknown samples (Z_{eff}). Later, Z_{eff} is estimated for each sample, and the dark matrix is defined as composed of a combination of two elements producing such an effective atomic number. This hypothetical composition is later supplied to the quantification program as "known dark" matrix. In Figure 3.16, a graphical view of this procedure is shown. For instance, this approach has been successfully applied for metal determination in vegetation samples (Marguí et al., 2006).

- **Influence coefficient algorithms**

Another type of mathematical method computes the matrix effect by means of coefficients, rather than by estimating the FP equation for each unknown. These coefficients can be determined theoretically or empirically and they can be based on concentrations or intensities (see schematic in Figure 3.17).

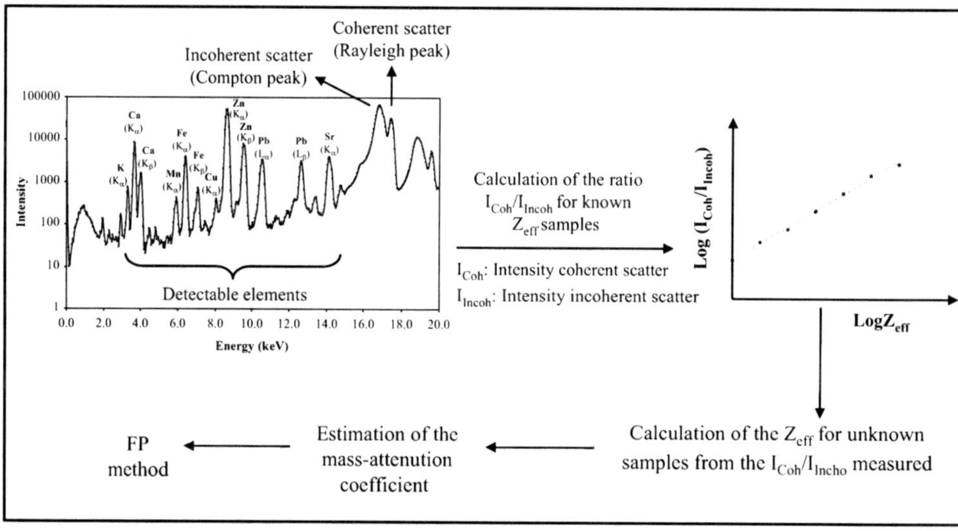

Figure 3.16. Schematic of the fundamentals of the BFP method.

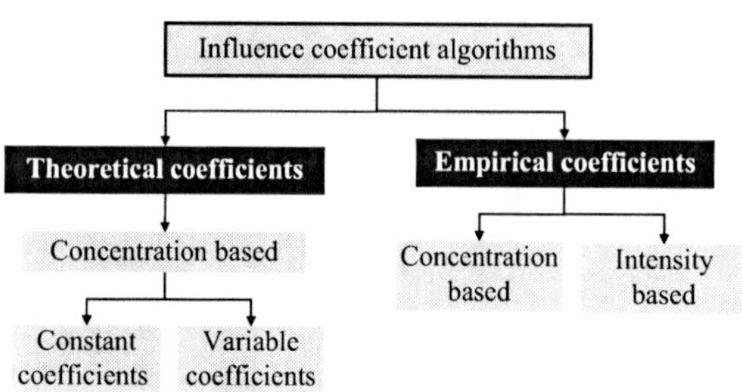

Figure 3.17. Broad subdivisions of application of interelement matrix correction algorithms.

In general, these models have the following expression:

- Concentration-based algorithms

$$W_i = K_i I_i \underbrace{\left(1 + \sum_j \alpha_{ij} W_j\right)}_{\text{Matrix correction term}} + B_i \quad \begin{cases} W_i = \text{Analyte concentration} \\ K_i = \text{Sensitivity} \\ I_i = \text{Analyte intensity} \\ \alpha_{ij} = \text{Influence coefficient} \\ W_j = \text{Concentration of matrix element} \\ B_i = \text{Analyte background} \end{cases} \tag{3.5}$$

- Intensity-based algorithms

$$W_i = K_i I_i \underbrace{\left(1 + \sum_j \alpha_{ij} I_j\right)}_{\text{Matrix correction term}} + B_i \quad \begin{cases} W_i = \text{Analyte concentration} \\ K_i = \text{Sensitivity} \\ I_i = \text{Analyte intensity} \\ \alpha_{ij} = \text{Influence coefficient} \\ I_j = \text{Matrix element intensity} \\ B_i = \text{Analyte background} \end{cases} \tag{3.6}$$

The influence coefficients depend on the analyte element and analyte line, the matrix elements and their concentration, the energy of the incident radiation, and the geometry of the system.

Once the coefficients have been calculated theoretically (by means of FPs) or empirically (using regression analysis), it is then possible to estimate the matrix correction term by the measurement of several standards containing known element weight fractions. Then, a regression line may be determined by plotting the measured intensity of the analyte multiplied by the corresponding matrix correction term against the analyte weight fraction. The weight fraction of the analytes in the unknown samples is determined by solving the set of equations with the general form shown in Figure 3.18, either algebraically or by means of an iterative method.

In Figure 3.19, an example of the benefit of using a correction method for matrix effects based on the use of influence coefficients is shown.

Depending on whether a single coefficient is considered per interfering element, or more than one is used, the influence coefficient is a constant for a given application, or the value of the coefficient varies with composition, different algorithms have been developed to calculate them for each case. These models are commonly named after the person(s) who proposed them and are accessible in commercially available software such as Lucas-Tooth and Price, Lachance and Traill, Claisse and Quintin, de Jongh, Rasberry and Heinrich, Lachance and Claisse, COLA (COmprehensive LAhance), Broll and Tertian, and Rousseau.

Usually, algorithms based on constant influence coefficients (i.e., predicting the effect of one element on the intensity/concentration of another in a given matrix) are only useful when analyzing thick samples, with monochromatic excitation and when enhancement effects are almost negligible. However, good quantitative results can be obtained for polychromatic excitation and/or when enhancement effects are present if the concentration range is limited (usually between 0% and 10%). Otherwise, the use of algorithms with variable coefficients (binary or multielement), which vary explicitly with concentration of one or more components in the

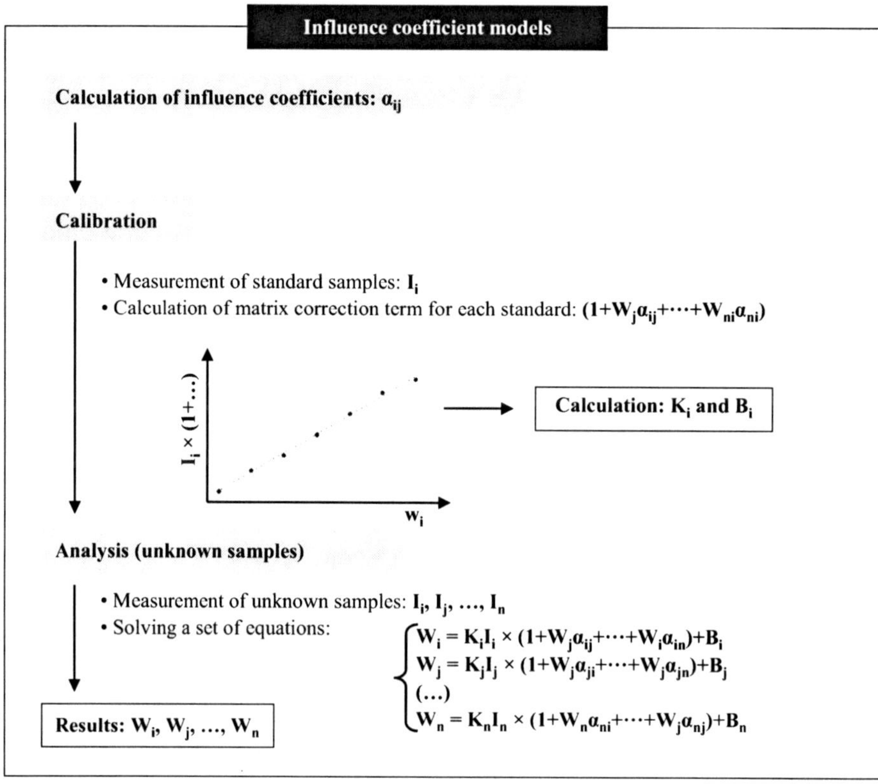

Figure 3.18. Influence coefficient model steps.

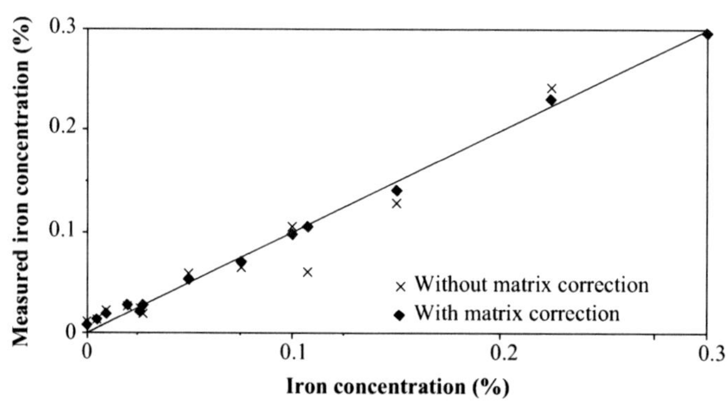

Figure 3.19. Example of the benefit of using a correction method for matrix effects based on the use of influence coefficients (referred to as intensity correction) (Fe determination in samples of vegetal origin) (Marguí, Hidalgo, & Queralt, 2005b).

specimen, becomes necessary to obtain quantitative results. These algorithms can be applied over a wider concentration range: up to 40% for binary variable coefficients, or up to 100% if multielement coefficients are used.

In Table 3.3, the concentrations determined using different matrix correction algorithms, both theoretical and empirical influence coefficients, for a specimen containing 48% Fe, 28% Cr, and 24% Ni are displayed.

Table 3.3. Comparison of concentrations calculated from different matrix correction algorithms using theoretical and empirical influence coefficients for a specimen containing 48% Fe, 28% Cr, and 24% Ni (Willis & Lachance, 2004)

Algorithm	Characteristics	Fe (%)	Cr (%)	Ni (%)
Theoretical algorithms				
• Claisse–Quintin	Concentration, variable	48.02	28.01	24.02
• COLA	Concentration, variable	47.92	28.07	23.99
• Lachance–Claisse	Concentration, variable	48.00	28.00	23.99
• Broll–Tertain	Concentration, variable	47.99	28.00	23.99
• Rousseau	Concentration, variable	48.00	28.00	23.99
Empirical algorithms				
• Lucas–Tooth and Price	Intensity, constant	47.74	27.33	21.47
• Rasbery–Heinrich	Concentration, constant	48.00	26.96	23.96
• Lachance–Traill	Concentration, constant	47.77	27.17	24.09

As is seen in this case, similar results were obtained using the different algorithms. Therefore, if any of the algorithms is used correctly, each will give acceptable results. It is up to the analyst to decide which particular algorithm is better to be used in each particular application. However, in many cases, the decision has to be based on the availability, quality, flexibility, and cost of the software used to apply a particular algorithm.

3.3.2.3 Overview of Correction and Compensation Methods

At this point, the basic question from a practical point of view is "What compensation or matrix correction method should we use?" There is not a unique answer to this question, as the choice of which method to use has to be determined by the particular application (type of sample to be analyzed, availability of standard samples, method of sample preparation, single element or multielement analysis, etc.). For instance, compensation methods will only correct for the effect of an unknown, but constant, matrix. Also, these methods do not require the analysis of all constituents in the specimen but usually require an additional sample preparation step (dilution, addition of an internal standard). They are usually useful when only one or few elements have to be quantitatively determined in an unknown but relatively constant matrix.

On the other hand, mathematical methods (FPs as well as methods based on influence coefficients) can handle cases in which the matrix effect is more variable from one specimen to another, and they are more appropriate if multielemental analysis is required. Therefore, they are more flexible than compensation methods but they require a wide knowledge of the matrix. Commonly, FP methods are more used in research laboratories and influence coefficient algorithms tend to be used for routine analysis in industry. A reason for that is based on considerations such as the availability of calibration standards and computer software, rather than on differences in analytical capabilities.

However, the final choice has to be made by the analyst.

3.3.3 QUALITY OF XRF ANALYTICAL RESULTS

One of the main tasks of any analytical chemistry laboratory is to provide reliable results on elemental concentration determination. Therefore, usually, an evaluation of the quality of the obtained XRF results has to be performed. In this context, a study of the limits of detection and quantification, linearity and range, precision, and accuracy is usually carried out in quantitative XRF analysis. Moreover, in order to check the quality of the XRF measurements over time or to correct for long-term drift of the XRF system, quality control specimens and drift-correction monitors have to be regularly analyzed. In this section, some recommendations are made to evaluate and ensure that quality XRF results are achieved.

3.3.3.1 Limits of Detection (LOD) and Quantification (LOQ)

When a new XRF analytical program is developed, the smallest concentration of the analytes that can be detected in a specimen has to be determined to check if the instrument is sensitive enough to detect all the elements of interest. The most current detection limit (LOD) used in XRF analysis is defined as the concentration resulting from a signal corresponding to 3 times the standard deviation of the background (with a confidence level equal to 99.86% according to the IUPAC recommendation (Currie, 1995). This procedure is also known as the 3-sigma criterion:

$$\mathrm{LOD} = \frac{3}{S_i}\sigma_\mathrm{b} \tag{3.7}$$

where S_i is the instrumental sensitivity and the σ_b is the standard deviation of the background. In general, the probability distribution of the results of a series of measurements for XRF signals can be considered as close to a Poisson distribution, and in such cases $\sigma_\mathrm{b} = \sqrt{b}$.

With WDXRF systems, it is a common practice to measure the intensity at the peak of a line and at a few background positions close the peak. Therefore, the background under the peak is determined by interpolating the intensities measured at the background positions. However, in EDXRF systems, the analytical signal is the peak area and the background is calculated from the continuum under the peak. This fact leads to different expressions for calculating the LOD, depending on whether a WDXRF or an EDXRF system is used (see Figure 3.20).

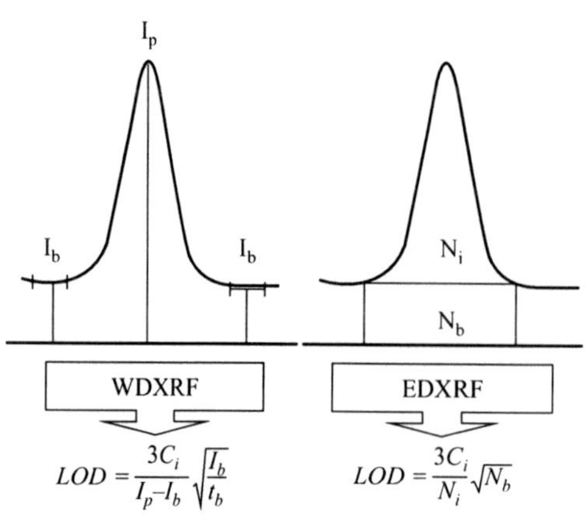

Figure 3.20. LOD evaluations for WDXRF and EDXRF systems.

Where C_i is the analyte concentration, I_p is the peak height, I_b is the background height, t_b is the time of the background measurement, N_i is the peak area, and N_b is the continuum under the peak.

However, in both cases, it is important to highlight that the LOD is not the smallest concentration that can be measured in a given analytical context but only of a theoretical estimation of the minimum concentration that can be detected based on the discrimination between the analyte signal and the background. For instance, this expression does not take into account the contribution for any instrumental or sample background. For this reason, a quantitative analysis is not possible at the LOD level (Rousseau, 2001).

For this reason, in quantitative applications, in addition to the LOD, it is also necessary to determine the smallest concentration of the analytes that can be quantified (not only detected) in a specimen. The most commonly used expression to evaluate the limit of quantification is based on the so-called 10-sigma criterion:

$$\mathrm{LOD} = \frac{10}{S_i}\sigma_\mathrm{b} \tag{3.8}$$

where S_i is the instrumental sensitivity and the σ_b is the standard deviation of the background. In general, the probability distribution of the results of a series of measurements for XRF signals can be considered as close to a Poisson distribution, and in such cases $\sigma_\mathrm{b} = \sqrt{b}$.

From aforementioned equations, it can be deduced that the LOD is better when the background is less, the instrumental sensitivity increases, and through longer counting times. As both the spectrometer sensitivity and the measured background vary with the average atomic number of a sample, the LODs will, from a practical point of view, vary with

- specimen matrix composition;
- analyte atomic number.

In general, for a given analytical context, the LOD will be smaller when the average atomic number of the matrix becomes lower.

This fact can be explained because the degree of absorption is lower for light element matrices and, therefore, the measured signal is higher. An example of this tendency is shown in Table 3.4.

As already stated in the introductory section, instrumental sensitivity falls dramatically toward the low energy limit (low-Z elements) owing mainly to low fluorescence yields and the increased influence of absorption. As a result, the poorer LODs for elements such as F and Na are at the levels of hundredths of one percent rather than the parts per million levels associated with higher Z elements.

3.3.3.2 Working Range and Linearity

In an XRF analysis, the working range and linearity of the method are defined by the selected correction or compensation method. For example, the use of normalization of the measured fluorescent to scattered radiation to compensate for differences in attenuation is valid only if there are no absorption edges of major elements between the analytical wavelength and the wavelength of the scattered radiation. In practice, this correction cannot be applied if the mass

Table 3.4. Limits of detection for low atomic number elements in different matrices (Modified from Rousseau (2001))

Element	Matrix	LOD (mg/kg)
Na	Aluminum oxide	395
Mg	Limestone	79
Mg	Al	105
Si	Steel	29
Si	Limestone	21
P	Oil	6
P	Nylon	2.4
S	Oil	1.4

fraction of any element with atomic number larger than that of the analyte is expected to be in the weight percentage range. Another example is the limited working range (usually from 0% to 10%) when using constant influence coefficients for correction of matrix effects.

Anyway, in all cases, it is necessary to use appropriate standards to find a relationship between the measured analytical signal and the analyte concentration. The number of standards used for quantification purposes, again, depends on the method selected for matrix effect correction. For instance, when using the FP method, the calibration process can be carried out using only a few standards, whereas when using influence coefficients for matrix correction a large set of standards is usually required.

To get appropriate quantitative results, standards have to be similar to the unknown samples (approximately the same matrix composition) and they have to be presented to the spectrometer in a reproducible and identical way to the target samples. Standards may be a certified reference material or a sample itself which has been precisely analyzed by another (reference) analytical technique. Today, there are 3,000 reference standards listed in commercial catalogs and hundreds more are in preparation. These standards include ores, metals and alloys, rocks, minerals, waste products, and dusts collected by electrostatic precipitators. Environmental problems have increased the demand for standards for wastes (mine dumps), industrial wastes, and isotopic materials. For more information on available reference materials, interested readers may refer to the website of the European Institute of Reference Materials and Measurements (IRMM; http://irmm.jrc.ec.europa.eu) or the U.S. National Institute of Standards and Technology (NIST; http://nist.gov/srm/).

However, for some applications it is difficult to acquire sufficient certified standards, with similar matrices to the samples, to achieve a good spread of data points over the range of each element to be determined. In such cases, a good option is the preparation of synthetic standards in the laboratory with a matrix similar to the target samples. For instance, the use of several synthetic cellulose standards to simulate the vegetal matrix proved to be a good option for the simultaneous determination of some macroelements (Na, Mg, Al, P, S, K, Ca), microelements (Mn, Fe, Co, Zn, As), and nonessential elements (Sr, Pb) in vegetation specimens (Marguí, Hidalgo, & Queralt, 2005b).

3.3.3.3 Precision and Accuracy

In quantitative XRF analysis, the overall uncertainty of an analytical result depends on the combination of errors introduced mostly by the sample preparation, the measurement of the analyte signal and the background, the calibration, and the corrections used for matrix effects. All these errors can be grouped into two main categories: the random errors which are related to the precision of the method and the systematic errors that are associated with the accuracy of the analytical results.

- **Precision**

The concept of precision according to the definition adopted by the International standard ISO/IEC/17025[1] is defined as "a measure of closeness of agreement between results." The estimation of precision (uncertainty resulting from random errors) is usually performed by calculating the standard deviation (SD) of n replicate measurements using the following expression:

$$\sigma_x = SD = \sqrt{\frac{\sum (x_i - \bar{x})^2}{n-1}} \tag{3.9}$$

where x_i is the value of each replicated measurement, $\bar{x}$ is the mean value of the n replicate measurements, and n is the number of replicates.

However, for comparison purposes, usually the relative standard deviation (RSD) is used:

$$RSD(\%) = \frac{SD}{\bar{x}} \times 100 \tag{3.10}$$

In XRF, the RSD values for replicate measurements of a single sample are supposed to reflect the uncertainty due to the instrument and counting statistics (repeatability). Usually, RSD values decrease when increasing the measuring time as is shown in Figure 3.21. Therefore, the counting statistical errors can be reduced to a minimum by selecting appropriate counting times.

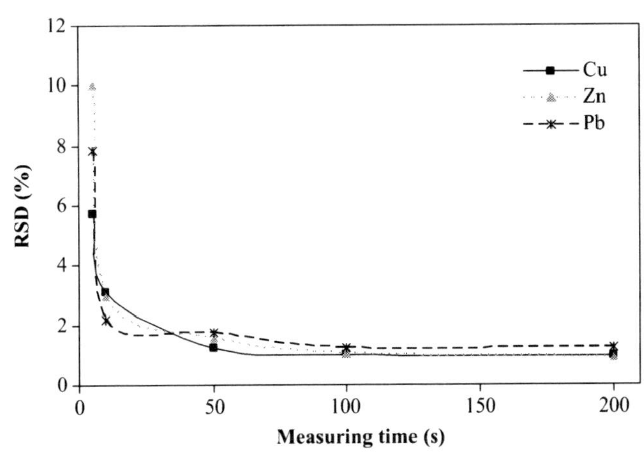

Figure 3.21. Improvement of repeatability in measurements by using different measuring times ($n = 5$, EDXRF system).

[1] http://www.iso.org/iso/catalogue_detail.htm?csnumber=39883; accessed October 2012.

If replicate preparations of a sample are measured, the uncertainty due to deviations in sample preparation has also to be taken into account. Then, according to the variance models ($\sigma_{total}^2 = \sigma_1^2 + \sigma_2^2 + \sigma_2^3 + \cdots + \sigma_n^2$), the influence of the sample preparation on the uncertainty can be estimated using the following expression:

$$\sigma_{total}^2 = \sigma_{Sample\ preparation}^2 + \sigma_{XRF\ Measurement}^2 \quad (3.11)$$

It can be assumed that the total variance is given by the results of the measurements of different sample replicates, as it is subjected to both the influence of the sample preparation and to the influence of the XRF measurements. As a numerical example, Table 3.5 compares the results of the measurement on six different pellet specimens to six measurements on the same pellet. As shown, in this case, the uncertainty is mainly due to sample preparation and not instrumental errors or counting statistics.

If the measurements are carried out on different days and by different operators, a better overall estimation of uncertainty is achieved. Therefore, an experiment designed to reflect the uncertainty of the method is supposed to comprise all the sources of random errors during analytical performance. In practice, the precision of the method can be improved by controlling the random errors introduced during sample preparation and by the analytical instrument.

- **Accuracy**

The concept of accuracy (trueness) according to the definition adopted by the International standard ISO/IEC/17025 is defined as "a measure of closeness of agreement between the arithmetic mean of a large number of test results and the true or accepted reference value." The uncertainty due to systematic bias resulting from the quantification model, matrix effects, or the presence of an instrumental blank signal can be estimated from the standard deviation of the bias of the results (q_i). Bias is defined as the difference between the obtained result, (X_i), and the certified value, X_{cert}, by using the following expression:

$$q_i = X_i - X_{cert} \quad (3.12)$$

Table 3.5. Comparison of the Pd results (in mg/kg) of one measurement on six different pharmaceutical pellet specimens and of six measurements on the same pharmaceutical pellet specimen. (Data taken from Marguí et al. (2009a))

Replicates	6 Measurements/Sample	1 Measurement/6 Sample replicates
1	18.3	20.4
2	18.4	18.4
3	18.5	17.5
4	18.4	18.3
5	18.6	18.3
6	18.4	20.5
Average	18.4	18.9
SD	0.11	1.23
RSD (%)	**0.58**	**6.5**

However, for comparison purposes, in some applications the relative bias in percentage (recovery) is used:

$$q_i = \frac{/X_i - X_{cert}/}{X_{cert}} \times 100 \quad (3.13)$$

In practice, to evaluate the accuracy of determined XRF results, the analysis of a certified reference material similar to the target samples that has not been utilized in the calibration is carried out. In some applications, it is difficult to get a suitable reference material with a matrix similar to the samples and then accuracy is evaluated by the analysis of spiked samples with the analyte of interest or by comparing the obtained results with a reference method. As an example, in Table 3.6, the results obtained for the analysis of a vegetation certified reference material using a WDXRF method (based on the use of influence coefficients) are displayed. As shown, good agreement was achieved between the certified values and data obtained by the WDXRF method, with recoveries ranging from 94% to 107%.

The accuracy of the XRF measurements, as already stated, depends mostly on the inaccuracies present in the quantification procedure. The FP method is often considered to be less accurate than using influence coefficient algorithms. This is caused primarily by the fact that the FP method is usually used with only a few standards. However, the accuracy of FP method can be very similar to that of influence coefficient algorithm when the same standards are used in both cases (many standards similar to the unknown).

Table 3.6. Concentrations obtained for the certified reference material GBW07602 (Bush, branches, and leaves) by using a WDXRF quantitative method based on the use of influence coefficients (results are the mean of three replicates, adapted from Marguí, Hidalgo, and Queralt (2005b))

Element	Certified	WDXRF method
Na (%)	1.10 ± 0.06	1.10 ± 0.09
Mg (%)	0.287 ± 0.011	0.272 ± 0.008
Al (%)	0.214 ± 0.018	0.20± 0.01
S (%)	0.32 ± 0.02	0.31 ± 0.01
K (%)	0.85 ± 0.03	0.94 ± 0.03
Ca (%)	2.22 ± 0.07	2.27 ± 0.05
Mn (mg/kg)	58 ± 3	54 ± 3
Fe (mg/kg)	1020 ± 40	1030 ± 60
Co (mg/kg)	0.39 ± 0.03	<0.5
Zn (mg/kg)	20.6 ± 1.0	22 ± 1
As (mg/kg)	0.95 ± 0.08	<1
Sr (mg/kg)	345 ± 7	310 ± 10
Pb (mg/kg)	7.1 ± 0.7	7.9 ± 0.7

Table 3.7. Evaluation of precision, accuracy, and combined uncertainty for an EDXRF methodology developed for plant analysis (adapted from Marguí et al. (2006)).

Element	Certified (mg/kg)	Average (mg/kg)	Relative precision (u_{prec})	Relative bias	Relative uncertainty due to bias (u_{bias})	Combined uncertainty (u_{comb})
Fe	264 ± 10	205	0.01	–0.22	0.007	0.01
	274 ± 10	218	0.01	–0.21	0.009	0.02
	n.c.	302	0.03	—	—	—
	1070 ± 40	940	0.004	–0.12	0.003	0.005
Cu	6.6 ± 0.4	9.8	0.01	0.48	0.02	0.02
	9.3 ± 0.5	10.5	0.02	0.12	0.02	0.03
	12.2 ± 0.6	17.9	0.03	0.47	0.04	0.05
	17.3 ± 1.0	17.0	0.01	–0.02	0.01	0.02
Zn	26.3 ± 0.9	20.6	0.02	–0.22	0.01	0.02
	37 ± 1	30.0	0.01	–0.19	0.01	0.02
	55 ± 2	50.4	0.02	–0.08	0.02	0.02
	82 ± 3	82.1	0.03	0.0007	0.03	0.04
As	0.068 ± 0.012	<0.2	—	—	—	—
	0.28 ± 0.03	0.36	0.10	0.28	0.31	0.33
	0.37 ± 0.06	0.34	0.25	–0.09	0.23	0.34
	1.25 ± 0.10	1.21	0.08	–0.03	0.08	0.11
Pb	(0.20) i.v.	<0.27	—	—	—	—
	1.5 ± 0.2	1.56	0.08	0.04	0.09	0.12
	4.4 ± 0.2	3.17	0.01	–0.28	0.007	0.01
	47 ± 2	47.7	0.02	0.02	0.02	0.03

n.c, not certified; i.v, information value

Once the uncertainty due to the random errors (precision) and the systematic errors (accuracy) have been determined, we can evaluate the "combined uncertainty" of the methodology (Padilla Álvarez et al., 2007). The calculation of this uncertainty is based on the law of the error propagation:

$$u_{comb} = \sqrt{(u_{prec})^2 + (u_{bias})^2} \tag{3.14}$$

where u_{comb} represents the combined uncertainty, u_{prec} is the uncertainty due to precision, and u_{bias} is the uncertainty due to systematic bias. In Table 3.7, an example of the calculation of the precision, accuracy, and combined uncertainty for an EDXRF methodology developed for plant analysis is shown.

Expressions used in the table:

$$U_{prec} = \frac{SD}{\bar{x}},$$

$$\text{Relative bias} = \frac{X_i - X_{\text{cert}}}{X_{\text{cert}}},$$

$$U_{\text{bias}} = SD\left(\frac{X_i - X_{\text{cert}}}{X_{\text{cert}}}\right),$$

$$U_{\text{comb}} = \sqrt{\left(U_{\text{prec}}\right)^2 + \left(U_{\text{bias}}\right)^2}$$

As can be seen from this table, for the case of Cu and Zn, the EDXRF determination was affected by the inaccurate fitting of the peaks (Cu-K_α = 8.047 keV, Zn-K_α = 8.638 keV) in the vicinity of an instrumental interference of W-L_α = 8.396 keV, leading to a larger relative bias in the determination of these elements. For Fe, in the case of samples containing high amounts of Mn, the authors found also a less accurate determination.

3.3.3.4 Quality Control of the Results

In addition to the analysis of reference materials to monitor trueness and the analysis of replicate samples to monitor precision, the analysis of blank samples is also recommended to account for possible contamination effects, and any periodic verification of the spectrometer to ensure the quality and reproducibility of the XRF measurements over time. For instance, for this latter case, the analysis of drift correction monitors should be regularly performed to correct for any long-term drift of the equipment. Correction for the drift can be made by measuring selected specimens for each of the analytes and by calculating the ratio between the observed intensities and those obtained when the calibration was performed. The ratios can then be applied as a correction to the drift. However, sometimes, drift correction does not bring the quality control specimens back into line with the expectations. In those cases, a recalibration process becomes necessary. The goal of the recalibration is to reconstruct the calibration graph but using only a few of the standards used for the initial calibration. The only requirement is that the recalibration specimens have count rates different enough to determine accurately the slope of the calibration curve.

CHAPTER 4

Sample Preparation Procedures

4.1 INTRODUCTION

The objective of an XRF analysis is to obtain accurate analytical data about the elemental composition of a specific material. This cannot be achieved unless the specimen is properly prepared. There is no doubt that the most common error source during an XRF analysis is inadequate and/or inappropriate preparation of the specimen and thus, the analyst must know how to optimally prepare the target specimen for each particular case. Several principles with respect to good sample preparation have to be taken into account, these are:

- **Adequately prepared specimen must be representative of the material.** Sample preparation usually involves taking an aliquot from a submitted sample, then preparing a specimen for analysis from this aliquot in such a way as to ensure that the relatively small volume of sample actually analyzed by the X-ray beam is representative of the total specimen. If the analyzed volume of the prepared specimen and the sample are of the same composition, only then can the analysis be considered meaningful.
- **Before any sample preparation method is selected, one should think of the purpose of the analysis (qualitative or quantitative analysis).** If only a qualitative analysis is required, usually, samples can be measured directly or with a minimum sample preparation. However, if a quantitative analysis has to be carried out then a previous sample treatment step becomes necessary to meet the conditions for homogeneity and a flat and smooth specimen surface. Moreover, for quantitative analysis, unknown samples must be presented for analysis in a reproducible and identical manner as to the standards employed for quantification calibration purposes.
- **Sample preparation will affect the final attenuation correction to be considered to compensate for sample self-absorption.** As commented in Section 3.3 (quantitative XRF analysis), the thickness of the measured sample can influence the final analytical signal. For this reason, when possible, the best option for quantitative analysis is to prepare the sample as a thin layer, as in such a case it is possible to consider that the total mass-attenuation coefficient of the sample is almost negligible and a linear relationship

between the intensity of the fluorescent radiation and the concentration can be assumed. However, as stated earlier, in most cases this is not possible and samples must be prepared as thick or intermediate thickness samples (for which cases the analytical signal becomes dependent on the sample thickness), thus, appropriate correction methods have to be used to account for this.

- **Specimen must fit into the specimen chamber of the XRF spectrometer.** Most spectrometers are designed to measure circular disks with a radius of 5–50 mm. Therefore, it is desirable to prepare the solid specimen in the shape of a round, flat disk. Another approach is to place the samples (liquid or powder) into cell cups that also fit into the specimen chamber of the XRF system.

  In systems that work in open air conditions, it is not necessary to place the sample to be analyzed into a cup of fixed external dimensions into a radiation protection chamber. Therefore, these systems have no constraints with respect to the dimensions and shape of the samples.
- **Specimen preparation should be avoided as much as possible.** Specimen preparation usually involves the addition of chemical reagents to the sample and/or to use crushing, pulverizing, and milling procedures (in the case of heterogeneous solid samples) to obtain sample particles that are of uniform and appropriate size to yield chemical sample homogeneity. In all cases, one has to be aware of the introduction of contaminants in these processes to the target sample. For instance, the use of an agate mortar can introduce traces of SiO_2, Mg, and Ca to the specimen. This might be of less concern for geological materials, but for biological materials these contaminants represent a major source of error. For this reason, specimen preparation procedures should be avoided or minimized as much as possible to avoid potential contamination sources.

4.2 GENERAL SAMPLE PREPARATION PROCEDURES

A wide variety of sample types may be analyzed by XRF, including solids, liquids, and even gases. Therefore, a wide variety of sample preparation methodologies have been developed for this purpose. The published literature on this topic is rather extensive and a more complete description of the sample preparation procedures may be found in some earlier books (Buhrke, Jenkins, and Smith, 1998; Marguí, Queralt, and Van Grieken, 2009; Van Grieken and Markowicz, 2002) and reviews (Marguí et al., 2010b; Van Grieken, 1982). In this chapter, only the most common procedures and some specific sample procedures used in several recent applications and developments are summarized (see Figure 4.1). Specialized sample preparation strategies for TXRF and spatial resolution analysis will be covered, more extensively, in Chapters 6 and 7.

4.2.1 SOLID SAMPLES

Determination of the composition of solid samples, without any sample preparation, is possible for samples that are homogeneous in all three dimensions and have a flat surface. However, the majority of solid materials require a previous sample pretreatment step before they can be analyzed in order to meet the requirement for homogeneity and to ensure quality and reproducibility of the measurements.

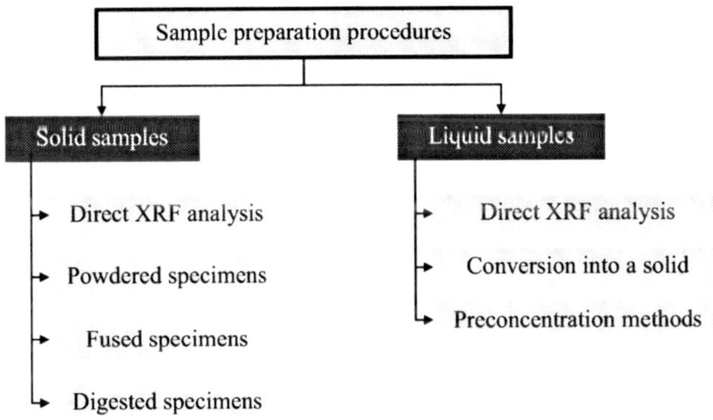

Figure 4.1. General XRF sample preparation procedures for solid and liquid specimen.

4.2.1.1 Direct XRF Analysis

Direct XRF analysis of solid samples, without any sample treatment, has proven to be very useful for chemically characterizing cultural heritage materials (e.g., ancient coins, study of oil paintings, ancient sculptures, and so forth) that are usually fragile and have a high potential for alterability if handled. (But a caveat is that one still has to be aware of possible surface patinas affecting an examination). Qualitative information derived from such an XRF analysis (usually accomplished nowadays using a handheld XRF unit) is of considerable interest for the conservation and provenance studies of these materials. Direct XRF measurements can also be performed in authenticity studies related to jewelry and gemstones. For instance, by means of a qualitative XRF analysis of a gemstone, it is possible to discriminate between a diamond (C) and its main "competitor" synthetic gemstone (cubic zirconia, ZrO_2). Therefore, a direct XRF analysis can be useful to gain qualitative information on precious samples that cannot be analyzed by using atomic spectroscopic techniques which normally require a previous digestion procedure of solid samples before analysis.

One can also obtain quantitative analytical information with a minimum sample preparation procedure in the study of metals and alloys, which usually entails only a surface preparation before the XRF analysis. The most suitable surface preparation methodology depends on the individual characteristics of the metal or alloy. The major surface preparation methods for these types of samples are polishing, etching, machining (milling, turning), or mechanical grinding. In recent years, the development and commercialization of small sample spot ED instrumentation (especially benchtop spectrometers, which offer extreme simplicity of operation in a low-cost compact design) have promoted the application of such instrumentation in the metal–alloy industry to carry out layer thickness measurements in a nondestructive way (without sample preparation).

4.2.1.2 Powdered Specimen

If the original solid sample is too heterogeneous for a direct analysis or it is not possible to obtain a suitable surface finish, then samples may be prepared as solid powders. The XRF analysis of powder samples ideally requires particles of an uniform size range and with a small enough grain size to yield chemical homogeneity without absorption problems. Reduction of the sample particle size is usually accomplished by a grinding procedure. For a routine XRF analysis, a particle size less than 62 μm is commonly accepted as appropriate. Various methods for grinding (manual/automatic) with grinding elements made of different materials (agate, corundum, stainless steel, and so on) are available. It is of importance to carefully select the type of grinding device for a particular application in order to avoid any contamination of the target sample.

Solid powder samples can be measured either as loose powders, as powdered suspensions, or as pressed pellets depending on the requirements for analytical accuracy and element detection.

- **Loose powders**

The simplest approach to analyze a loose powder sample is the introduction of a few grams of material into a three-fourth full sample cup[1] which is in fact a metal or plastic cylinder with a very thin polymer foil at the bottom, without any additional preparation. Alternatively, if the amount of a powdered sample is very limited, in some applications, small amounts of the powder may be distributed onto an adhesive side of a tape or a very thin polymer foil. Owing to the looser packing of the powder and the cup foil's absorption, the sensitivity for the lighter elements, such as Na, is much lower compared to the analysis of pressed pellets. Other shortcomings are that the bulk density may not be consistent and grain size variation may cause the XRF readings to vary. Nevertheless, depending on the requirements for analytical accuracy and trace element detection, the "loose" powder preparation can be a rapid alternative, convenient method compared to the pressing of the materials.

- **Pelletizing**

To reduce surface effects and to improve the precision of the analytical results, powders can be prepared as pressed pellets. Depending on the self-bonding properties of the powder, pellets are prepared directly by applying a high pressure to the powder in a die or by addition of a binder before pelletizing to form stable pellets and prevent cracking of the sample at the die surface. Binders are usually composed of light elements or organic materials such as cellulose, polyvinyl alcohol, boric acid, graphite, and so forth. At present, in the published literature, a wide variety of pelletizing methodologies are described which make use of different sample amounts of powder (from a few hundred milligrams to several grams), grain sizes, pressures, time of pelletizing, and so on. The final choice will depend on the aims of the analysis and characteristics of the sample.

- **Powdered suspensions**

In some applications, especially for water-insoluble materials, a powdered suspension can be prepared by mixing a few milligrams of powder with a few milliliters of water or another dispersant agent (i.e., a solution of a nonionic detergent). After the formation of a turbulent suspension, the sample is filtered through, for example, a Nuclepore™ filter; the resulting fairly uniform thin layer present on the filter is then measured directly by XRF. Another alternative is to deposit a few microliters of the resulting suspension onto a reflector for the subsequent TXRF analysis. Sample preparation details for TXRF analysis will be discussed in depth in Chapter 6.

---

[1] Breitländer Website (http://www.breitlander.com); Accessed November 2012.

4.2.1.3 Fused Specimen

Samples that tend to remain heterogeneous even after grinding and pelletizing are often treated by the technique of flux fusion. In principle, it involves fusion of the sample with an excess of flux (usually Na or Li tetraborate) and casting into a solid bead. Chemical reaction in the melt converts the phases present within the sample into glass-like borates, giving a homogeneous bead of dimensions suitable for direct placement into the spectrometer sample chamber. The choice of the ratio of sample-to-fusion mixture must be chosen with special care because it controls the speed and degree of the chemical reaction, the final mass absorption coefficient, and the dilution factor applied to the analytes. The latter case is of particular importance if trace element determination is required. Many useful recipes for fusions can be found in published literature on the topic (e.g., Buhrke, Jenkins, and Smith, 1998).

4.2.1.4 Digested Specimen

For some applications, the use of more complex sample treatment procedures, similar to those employed in atomic spectroscopic techniques, is required. In this context, dry and wet ashing procedures have been used to destroy the organic matter and dissolve the analytes of interest in the solid sample prior to XRF analysis. Dry ashing consists of calcination of the sample in a muffle furnace at temperatures between 450°C and 550°C. Then the resulting inorganic residue (ashes) is dissolved by an appropriate acid prior to the analysis. The main drawback of dry ashing is the inherent problem of volatilization losses of some elements (i.e., Hg, As, Se) during the ashing process. For this reason, wet ashing procedures, which involve a sample digestion with strong acids using lower temperatures, is preferred. The choice of the best decomposition procedure for target solid samples should be preceded by verification of the procedure for each specific matrix and analyte under study. For instance, a large number of metals and alloys can be brought into solution in aqua regia (a mixture of one part concentrated nitric acid with three parts concentrated hydrochloric acid) with gentle heating to 60°C. Although for the determination of total element content in geological matrices containing silicates, the addition of hydrofluoric acid to aqua regia is necessary.

When applying these sample preparation procedures, a liquid specimen is obtained. In the next section (Section 4.2.2), the treatments employed before XRF analysis of liquid samples are discussed.

4.2.2 LIQUID SAMPLES

Direct analysis of solutions can be performed on a few milliliters placed in a cup, usually made of polyethylene or poly(tetrafluoroethylene) (PTFE), with a thin film bottom in a nonevacuated XRF setup. The most common types of films used for such purpose are Mylar® (poly(ethylene terephthalate)), polypropylene, polycarbonate, polyimide (Kapton®), and Teflon® (PTFE). The preferred film choice is based on factors such as X-ray transmission, impurities, chemical resistance, and cost. Bubbles released from solutions due to inadequate sample holder filling and heating of the solution during an XRF measurement give rise to certain reproducibility problems when analyzing liquid samples. Moreover, liquid samples usually provide a high

X-ray scatter background, resulting in a poor signal-to-noise ratio with limits of detection in the milligrams-per-liter range, which are not satisfactory for most environmental and industrial applications.

The sensitivity can be improved using TXRF spectrometry (for more details, see Chapter 6) or performing a sample treatment before the XRF analysis. In this case and taking into account that solid specimen are more favorable for measuring by XRF than liquid specimen, some workers have proposed transforming liquids into quasisolid specimen by increasing their viscosity using thickening agents (e.g., gelatine, agar, starch, or water-soluble organosilicone polymers) (Eksperiandova, Makarovska, and Blank, 1998). Then, the compact specimen obtained conform to the shape of the cuvette in which they solidify and, thus, they are suitable for different X-ray spectrometers. Nevertheless, in some applications, more sophisticated sample treatments to remove the sample matrix and improve the limits of detection are necessary (preconcentration methods). In the next section, the most commonly used preconcentration procedures are presented.

4.2.2.1 Preconcentration Methods

Preconcentration is a process as a result of which the ratio of the concentration or the amount of a desired trace element to the macrocomponents (matrix) increases. Preconcentration methods are based on physical and/or chemical principles and, therefore, they are usually grouped into two main categories: physical preconcentration methods and chemical preconcentration methods (see Figure 4.2). In this section, a summary of the main characteristics of each procedure are presented. A comprehensive overview of this subject can be found in published review articles on preconcentration methods (Marguí et al., 2010b; Van Grieken, 1982).

It is interesting to note that no method of preconcentration will be a panacea because the optimum choice will be influenced by the types of liquid that have to be examined, the elements of interest, the number of samples, the required accuracy, precision and sensitivity, the available equipment, and so on.

- **Physical methods**

Freeze drying. Freeze drying is a dehydration process of the sample. It consists of rapid freezing followed by sublimation of the frozen water in the material directly from the solid phase

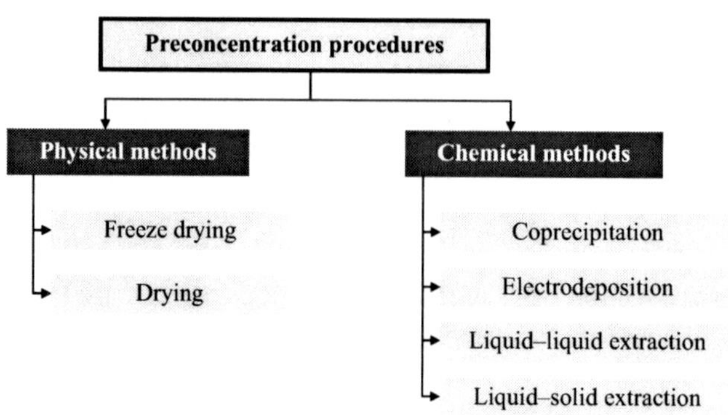

Figure 4.2. Most commonly used preconcentration methods for XRF analysis.

to gas. Then the evaporation residue can be pelletized and measured by XRF. With such a procedure, detection limits of a few micrograms per liter can be achieved. This preconcentration procedure has been most commonly applied for the analysis of biological samples (i.e., blood) or infusion and brewing of medicinal herbs.

Drying. A simple way to preconcentrate ions from liquid samples is to evaporate the solvent. The method involves depositing a known volume of a liquid sample, typically milliliter to microliter, onto a thin film substrate (Prolene™, Mylar, Kapton, filter paper, and so on) and allowing the liquid to dry into a concentrated dried residue or "dried spot." Unfortunately, the use of such methodology is limited to nonsaline samples because the formation of finite crystals during the drying procedure can lead to large and unknown X-ray absorption effects resulting in problematic matrix corrections. Recent advances covering the topic of drying preconcentration methods are focused mostly on the use of nanodroplets for dried spot preparation (coupled to μ-XRF or TXRF spectrometry). This method, developed by Miller and coworkers (Miller and Havrilla, 2004), is a variation of the typical dried spot technique in which instead of depositing milliliter or microliter volumes, nanoliter sample volumes are deposited onto substrates for X-ray analysis. Advantages of this procedure include a decreasing of drying time and reagent consumption and an increase of sample throughput. Moreover, the resulting dried spots are uniform in shape and elements contained in them can be detected at picogram detection limits. This new sample preparation method has the potential for use in situations where sample availability is limited. One of the major drawbacks of the current implementation of the nanoliter spot technique is sample preparation errors due to variation of contact of the injector tip geometry onto the substrate surface between droplet injections. In this sense, Miller and coworkers have reported in a recent paper the use of noncontact automated printing technology (APT) as an alternative method for nanoliter dried spot preparation (Miller, Hastings, and Havrilla, 2006).

- **Chemical methods**

Chemical preconcentration methods are based on chemical principles and they are more widely used compared to physical methods.

Coprecipitation. The basic principle of coprecipitation is the addition of a precipitating agent to the solution and the determination of the target elements in the solid obtained after a filtration procedure. With such an approach, detection limits at the low microgram per liter range can be obtained. A general schematic of this procedure is shown in Figure 4.3.

Because of the low aqueous solubility of their metal chelates, dithiocarbamates (sodium dibenzyldithiocarbamate (Na-DBDTC) or ammonium pyrrolidine-dithiocarbamate (APDC)) have been widely used as precipitating reagents. In some applications, before the addition of the precipitating agent, some additional treatments are performed including the addition of a suitable element to act as a carrier or coprecipitant, the elimination of organic matter by using oxidizing agents, or the chemical separation from major matrix elements by using cation exchange resins.

To collect the formed precipitate, different film substrates can be used including filter papers or thin polymer films (e.g., polypropylene, Mylar, Kapton). Usually, the thickness of the resulting precipitate can be considered as a thin film and, thus, it is not necessary to apply a correction of the matrix effect for further XRF analysis. The coprecipitation method also has some drawbacks, mostly associated with laborious and time-consuming system operation.

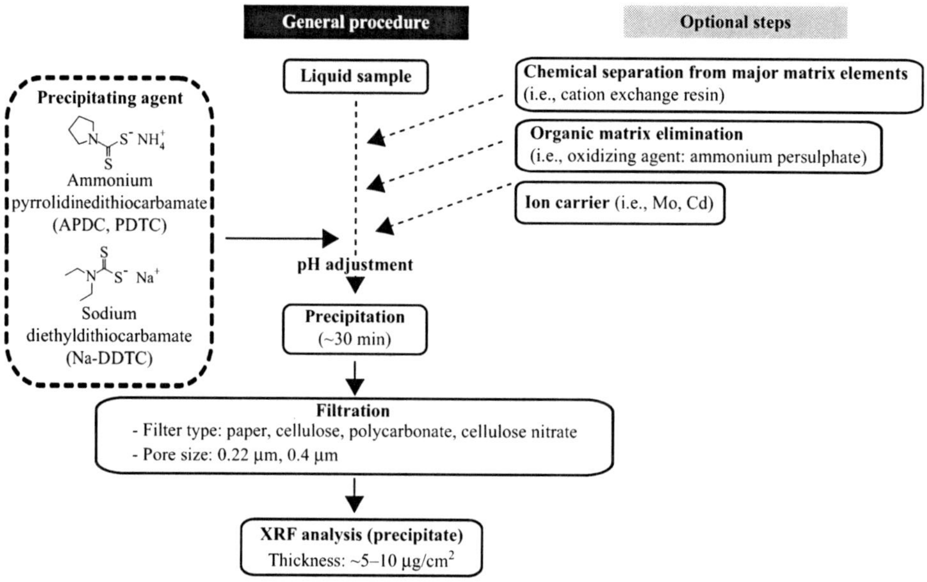

Figure 4.3. Coprecipitation procedure and subsequent XRF analysis (Modified from Marguí et al., 2010b).

Electrodeposition. This method uses electrical current to reduce elements from a solution and coat a conductive surface with a thin layer of the target element. After the electrolysis process, the electrode is removed from its holder, rinsed with water, and dried prior to XRF analysis. Despite the fact that the electroenrichment of elements onto a cathode seems to be a very interesting sampling approach for XRF analysis, its application has been limited owing to the necessity of sophisticated equipment compared to other preconcentration procedures.

Liquid–liquid extraction. This is a method to separate compounds based on their relative solubilities in two different immiscible liquids, usually water and an organic solvent. This procedure has been widely applied in atomic spectroscopy but its application combined with XRF spectrometry has been limited because, after the liquid–liquid extraction, an evaporation of the organic phase (containing the target analytes) is necessary prior to XRF analysis. Most studies using liquid–liquid extraction for subsequent XRF analysis are based on the complexation of elements under investigation using carbamates followed by solvent extraction with an organic reagent (i.e., chloroform or methylisobutylketone (MIBK)). Then, the water-immiscible liquid phase is collected and dried on a filter and analyzed directly by XRF. Another approach is to analyze a few microliters of the resulting organic phase directly by TXRF spectrometry (see Chapter 6).

Liquid–solid extraction. The basic principle of this procedure is a transfer of analytes from the aqueous phase to the active sites of an adjacent solid phase (using appropriate optimal conditions in the system liquid phase-analyte-sorbent) followed by a desorption step of the analytes with a suitable eluate. The use of a liquid–solid preconcentration step is a very favorable procedure to be combined with XRF as the solid phase is a convenient substrate for XRF measurements and thus, the elution step, required when atomic spectroscopic techniques are

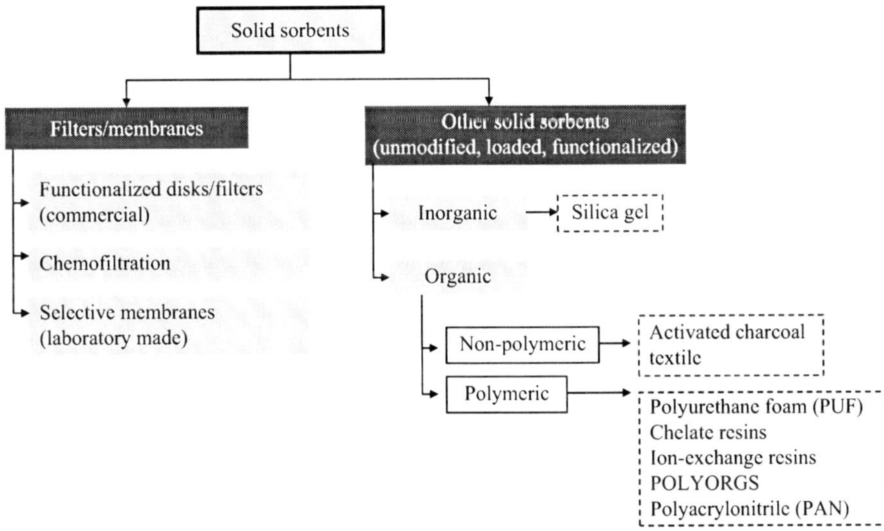

Figure 4.4. Solid sorbents used in solid-phase extraction for subsequent XRF analysis. POLYORGS, fibrous complexing sorbents containing amidoxime functional groups.

used, can be avoided. Solid sorbents with different characteristics and shapes can be used for such a purpose. In Figure 4.4, the most commonly used solid sorbents for solid-phase extraction and subsequent XRF analysis are shown.

As mentioned in Section 3.3, the ideal samples to be analyzed by XRF are thin-layer organic materials because of the reduction of the absorption-enhancement effects and the background of the exciting radiation scatter. For this reason, ion-collecting filters and extraction disks have been widely used as solid sorbents to preconcentrate metals for subsequent XRF measurements. An additional advantage of using these materials is the collection of the elements onto the solid by means of a simple filtration procedure. This system permits the use of a large volume of a sample (1L or more), resulting in large enrichment factors. Commonly, the disks and filters (made of cellulose, cellulose acetate, nitrocellulose, polystyrene, polyamides and nylon, among others) are functionalized with suitable chemical molecules (e.g., iminodiacetate group, diethylenetriaminetetraacetate, 8-hydroxyquinoleine) to allow for effective extraction of the target elements.

Another approach when functionalized filters or disks are unavailable is the complex formation of the elements to be determined with selective organic reagents in solution followed by sorption of this complex onto a nonselective membrane or filter in the course of filtration (Chemofiltration procedure).

The preparation of appropriate membranes containing extracting agents or various metal complexes is also a key way of producing suitable thin layers for preconcentration. The membrane preconcentration experiments are carried out by placing the aqueous solution into contact with the prepared membrane under continuous stirring for an optimized period of time. Then the loaded membrane is removed from the vessel, washed with deionized water, dried, and mounted directly into the spectrometer for XRF analysis. This preconcentration methodology has been successfully applied in combination with different configurations of XRF spectrometers for the

Table 4.1. Specific sample treatment procedures in some biological, environmental, and industrial application fields

	Type of sample	Sample preparation procedure
Biological	Vegetation	*Purpose: Quantitative analysis* - Oven-drying (T ≤ 60°C) or freeze-drying - Grinding to obtain a fine powder - Pelletizing (using different sample amounts/binders)
Biological	Biological tissues	*Purpose: Quantitative analysis* - Freeze-drying → pulverization → pelletizing *Purpose: Element distribution* - Freezing (liquid nitrogen) → cut into slices (microtome)
Biological	Body fluids (blood, serum)	*Purpose: Quantitative analysis* - Deposition of a few microliters onto a filter paper/thin film - In some applications a preconcentration step is needed
Environmental	Soil/sediment	*Purpose: Quantitative analysis* - Oven-drying - Grinding to obtain a fine powder - Pelletizing (using different sample amounts/binders) or fused bead preparation (determination of light and major elements)
Environmental	Aerosol and dust	*Purpose: Quantitative analysis* - Drawing a large air volume through a filter - Analysis of the loaded filter
Industrial	Automotive catalysts	*Purpose: Quantitative analysis of precious metals* - Grinding to obtain a fine powder - Pelletizing using wax as a binder
Industrial	Food (powder milk)	*Purpose: Quantitative analysis* - Grinding to obtain a fine powder - Pelletizing (using different sample amounts/binders)
Industrial	Pharmaceutical (solid)	*Purpose: Quantitative analysis* - Homogenization of the powder - Pelletizing (usually the addition of a binder is not necessary)
Industrial	WEEE*	*Purpose: Quantitative analysis of hazardous substances* - Method 1: Homogenization → pulverization → pelletizing - Method 2: Direct analysis → intelligent software

* Waste electrical and electronic equipment.

selective determination of trace amounts of target elements (e.g., Cd, Cr(VI), Pd(II), and Pt(IV) in complex liquid samples) (Marguí et al., 2008).

In addition to extraction disks, filters, and membranes, other solid sorbents have been used for liquid–solid extraction preconcentration prior to XRF sample analysis. In Figure 4.4, a classification of these sorbents based on their chemical nature (inorganic/organic) is shown.

Probably two of the most commonly used solid sorbents for metal preconcentration are synthetic resins (ion-exchange or polymeric) and silica gel (with and without functional groups). Low element contents in solutions can be concentrated with such sorbents and eluted and evaporated onto filter paper disks for subsequent XRF analysis. Alternatively, the sorbent itself can be used as a sample support either by pelletizing the solid or by depositing it onto a filter paper (through a filtration procedure) and covering the resulting specimen with a Mylar film for XRF measurements. Other sorbents such as polyurethane foams (PUF) or natural sorbents (charcoal) also offer a high capacity and a suitable sorption speed for chemical species in solution.

4.3 SPECIFIC SAMPLE PREPARATION PROCEDURES

We have earlier focused on the general sample treatment procedures for XRF analysis of solid and liquid samples. In this chapter, some specific sample preparation procedures commonly used in biological, environmental, and industrial fields are presented (see Table 4.1). XRF spectrometry has always been widely used in these areas owing to the inherent flexibility of the technique for the provision of qualitative and/or quantitative data in a laboratory and in situ environments. Detailed information on other specific sample preparation procedures may be found in published books (see Bibliography) specifically focused on sample preparation strategies (e.g., Buhrke, Jenkins, and Smith, 1998; Marguí, Queralt, and Van Grieken, 2009).

CHAPTER 5

Wavelength/Energy Dispersive X-Ray Fluorescence Spectrometry (WDXRF/EDXRF)

5.1 INTRODUCTION AND BASIC PRINCIPLES

Usually, XRF spectrometers may be categorized into two main types depending on the fundamentals of their detection system: wavelength dispersive systems (WDXRF) and energy dispersive systems (EDXRF). To understand the basic principles of each configuration it is important to recall the origins of commercial XRF spectrometers by referring back to the middle of the previous century.

The first modern commercial X-ray spectrometer was developed by Friedman and Birks in 1948. This device was a variation of a diffractometer for chemical analysis and it was composed of a W-target X-ray tube, a single crystal ((200)NaCl or (220)CaF_2), and a Geiger counter (particle detector that measures ionizing radiation) as a detection system. Therefore, the basic principle of this configuration was the use of diffraction by a dispersing system (the crystal) to separate the characteristic photons emerging from the sample in terms of their wavelength according to Bragg's law. This early spectrometer, named as a *wavelength dispersive X-ray spectrometer* (WDXRF), was able to provide qualitative and quantitative information for all elements above atomic number 22 (Ti) but the main limitation was the low intensity that the Geiger counter could tolerate without becoming saturated. With the advent of gas-proportional and scintillation detectors with a shorter dead time, counting at much higher rates became possible. Although the earliest commercial XRF devices operated under simple air path conditions, machines were soon developed utilizing He or vacuum paths, permitting the detection of lighter elements. In the 1960s, XRF devices began to use LiF crystals for diffraction and Cr and Rh target X-ray tubes to excite longer wavelengths. These developments were quickly followed by those of multichannel spectrometers for the simultaneous measurement of many elements and computer-controlled XRF devices.

However, perhaps the most significant development in XRF systems came in 1966 with the creation of the Li-drifted Si detector (Si(Li)) that provides very high resolution and X-ray photon separation without the use of an analyzing crystal. Therefore, from this moment forward, the manufacture of simpler and more economical spectrometers with only two basic units (excitation source and detector system) became possible. These new systems were known as *energy dispersive X-ray spectrometers* (EDXRF) and owing to their flexibility were even used on the Apollo 15 and 16 space missions in the early 1970s to determine the major elemental composition for atmosphere-free planetary surfaces.

Since then, noticeable development has been made in the instrumental aspects of XRF spectrometers. This progress includes, for instance, considerable improvements in the design and production of detectors and X-ray optics. This has resulted in a wide variety of instrumentation becoming available today since the first commercial X-ray spectrometers were developed. Within the scope of this chapter, WDXRF and EDXRF systems in which X-ray tubes are used for excitation will be considered. The material in this chapter discusses the instrumental aspects of current WDXRF and EDXRF systems and the main analytical capabilities of both configurations. In addition, some interesting applications and case studies for EDXRF and WDXRF spectrometers are reviewed at the end of this chapter.

5.2 WDXRF AND EDXRF LAYOUTS

In Chapter 2, the main types of sources, modifiers, and detector systems were considered in detail. In the next sections, Sections 5.2.1 and 5.2.2, only the spatial configuration of these devices in WDXRF and EDXRF systems will be considered. Therefore, for more detailed information on the basic components of each system, readers should refer back to Chapter 2.

5.2.1 *WDXRF INSTRUMENTATION*

The main characteristic of WDXRF systems is that the characteristic radiation emitted from the sample is separated according to its wavelengths using a diffraction device. The actual spectrometer itself consists of an X-ray tube, a specimen holder support, a primary collimator, a diffraction device, a second collimator, and a detector, all mounted on a goniometer. In some WDXRF systems, source modifiers such as primary filters are also placed between the X-ray source and the sample to improve the instrumental selectivity and sensitivity for specific element determination (for details, see Section 2.5 in Chapter 2). The geometric arrangement of the components in a WDXRF instrument is shown in Figure 5.1.

The X-ray source irradiates the sample and a portion of the characteristic fluorescent radiation from the specimen is passed via a collimator onto the surface of the diffraction device (analyzing crystal or a multilayer structure), where individual wavelengths are diffracted to the detector according to Bragg's law. A goniometer is used to maintain the required θ to 2θ relationship between the diffraction device and the detector. Owing to the high resolution power of single crystals or multilayered structures used for this purpose and the use of additional Soller (secondary) collimators, photons corresponding to characteristic lines with close energies can be detected without interfering with each other, thus providing a high specificity in the analysis. As spectral interference is avoided, the detection can be performed using proportional counters with large efficiency even at low energies.

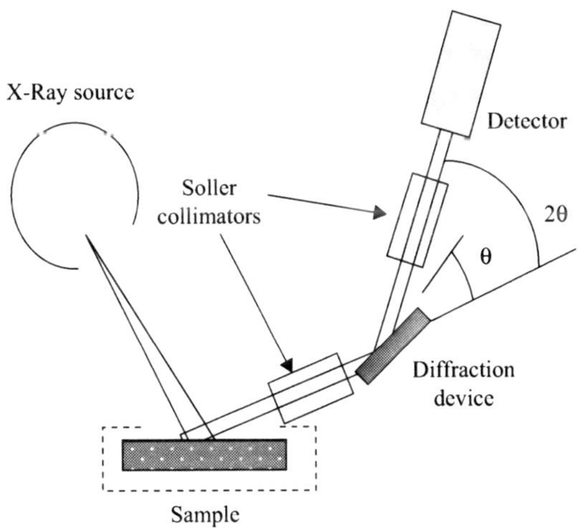

Figure 5.1. Geometric arrangement of a WDXRF system.

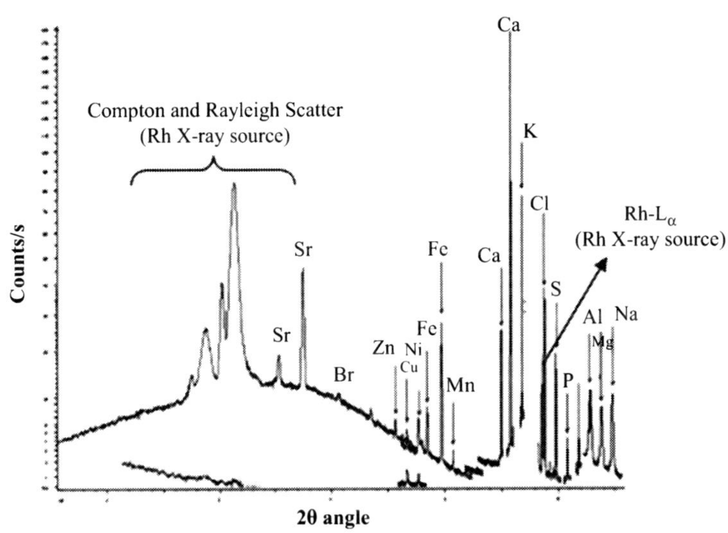

Figure 5.2. Basic spectrum recorded for the analysis of a leaf sample using a WDXRF spectrometer.

Usually, in WDXRF spectrometers, the analysis of different elements is carried out in a sequential way by synchronously changing the type and orientation of the monochromator device and the detector. In general, about six different dispersing devices are provided giving the analyst a wide range of choice of dispersion conditions. The smaller the *d*-spacing of the crystal, the better separation of the lines, but the wavelength range that can be covered becomes smaller. A tandem detector system (a gas flow counter and a scintillation counter) is typically employed to convert the diffracted characteristic radiation into voltage pulses that are integrated and displayed as a measure of the characteristic line intensity. As already stated in Table 2.3 (see Chapter 2), the gas flow counter is ideal for the measurement of longer wavelengths (light elements) and the scintillation counter is preferred for short wavelengths (medium-high elements). In Figure 5.2, a basic spectrum recorded for the analysis of a leaf sample using a WDXRF spectrometer equipped with a Rh X-ray tube is displayed.

Discontinuities observed in this spectrum result from the use of different analyzing crystals and excitation/detection conditions (see Table 5.1 for details).

Also available commercially are WDXRF multichannel spectrometers that are equipped with a set of fixed detection systems. Each detection system has its specific crystal and detector and each system measures the radiation of a specific element. The intensities are measured all at the same time; this is the reason that these spectrometers are called *simultaneous* systems. However, the number of diffraction devices/detectors sets is limited and usually simultaneous WDXRF instruments are only capable of determining 12–14 elements simultaneously.

The reader will find the actual addresses of several WDXRF manufacturers' headquarters in Chapter 9. In this context, it is interesting to remark that during the last decade, most of the WDXRF analyses have been performed using large-scale instruments with high-power X-ray tubes. However, at present, some of the manufacturers such as Rigaku Corporation offer bench-top WDXRF spectrometers equipped with air-cooled X-ray tubes that allow for the determination of low concentration levels (even for light elements) using a compact and less expensive system.

Table 5.1. Measuring conditions used to obtain the WDXRF spectrum in Figure 5.2

Z	Element	Line	Crystal[a]	Detector[b]	Collimator (in degrees)	Voltage (in kV)	Current (in mA)
11	Na	Kα	OVO-55	FPC	0.46	20	50
12	Mg	Kα	OVO-55	FPC	0.46	20	50
13	Al	Kα	PET	FPC	0.46	20	50
15	P	Kα	PET	FPC	0.46	20	50
16	S	Kα	PET	FPC	0.46	20	50
19	K	Kα	LiF200	FPC	0.46	50	20
20	Ca	Kα	LiF200	FPC	0.46	50	20
25	Mn	Kα	LiF200	SC	0.23	50	20
26	Fe	Kα	LiF200	SC	0.23	50	20
28	Ni	Kα	LiF200	SC	0.23	50	20
29	Cu	Kα	LiF200	SC	0.23	50	20
30	Zn	Kα	LiF200	SC	0.23	50	20
35	Br	Kα	LiF200	SC	0.23	50	20
38	Sr	Kα	LiF200	SC	0.23	50	20

[a] OVO-55, W/Si multilayer; PET, Pentaerythrite; LiF200, Lithium fluoride.

[b] FPC, Flow proportional counter; SC, Scintillator counter.

5.2.2 EDXRF INSTRUMENTATION

Unlike the WDXRF systems, EDXRF spectrometers consist of only two basic units: the excitation source and the detection system (see Figure 5.3).

For the EDXRF case, as the resolution of the energy dispersive system is equated directly to the resolution of the detector, typically a semiconductor detector of high intrinsic resolution

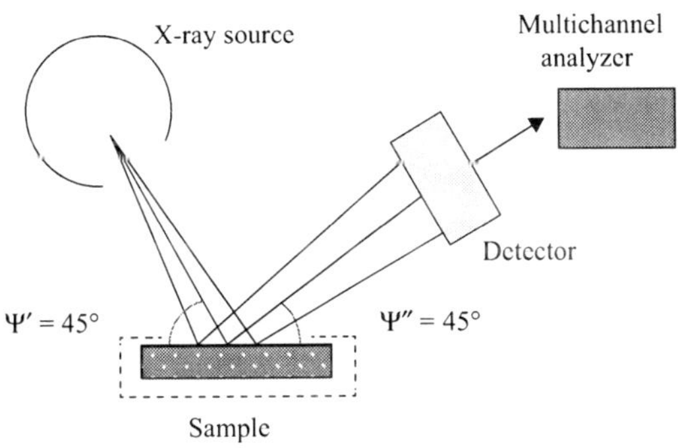

Figure 5.3. Schematic arrangement of a conventional EDXRF system.

is employed (Si(Li)). The use of this type of detector allows for recording an electronic signal, the current of which is proportional to the energy of the detected photon. A multichannel analyzer is used to collect, integrate, and display the resolved electronic pulses.

It is interesting to note that using this configuration, all of the X-rays emitted by the sample are collected at the same time (simultaneously measured), giving greater speed in the acquisition and display of data. In practice, however, there is a limit to the maximum count rate that the spectrometer can handle. For this reason, usually source modifiers such as filters are placed between the X-ray source and the sample to reduce the intensity of interfering lines and/or the continuum of the X-ray tube and thereby avoiding saturating the detector. The characteristics of the filter are defined by its X-ray absorption curve that is controlled by selecting the material and thickness (for more details see Table 2.4 in Chapter 2).

Another option is the introduction of a pure element or compound target between the primary source and the sample in such a way that a selectable energy range of secondary photons is incident upon the sample. This allows for selective excitation of certain portions of the energy range, thus increasing the fraction of analytically useful photons while decreasing the unwanted photons that enter the detector. While this configuration does not completely eliminate the count rate and resolution limitations of the primary system, it certainly reduces them. Commercial instruments usually offer a choice of several secondary targets. In current commercial instruments, using secondary target excitation, the primary, the secondary, and the characteristic radiation from the sample are configured at mutually orthogonal angles (3D geometry) as can be seen in Figure 5.4.

With this geometry, a significant reduction of the background in the fluorescence spectrum is achieved because the exciting radiation is polarized when scatter occurs through the right angle and cannot then be scattered a second time and reach the detector. Therefore, the sensitivity and limits of detection for minor and trace elements are better than those achieved with the older and more common 2D secondary target spectrometers (see, for example, the spectra of Figure 5.5).

Furthermore, novel polarized-beam EDXRF instrumentation combines the benefits of polarization (triaxial geometry) with the possibility of using different secondary targets to produce fluorescence lines to act as near-monochromatic secondary sources at slightly higher energies than the absorption edges of the analytical lines to be excited. Therefore, by combining polarization with the possibility of different secondary sources, an instrument can be designed to generate a range of excitation and polarization conditions optimized for different groups of analytes (see Figure 2.19 in Chapter 2).

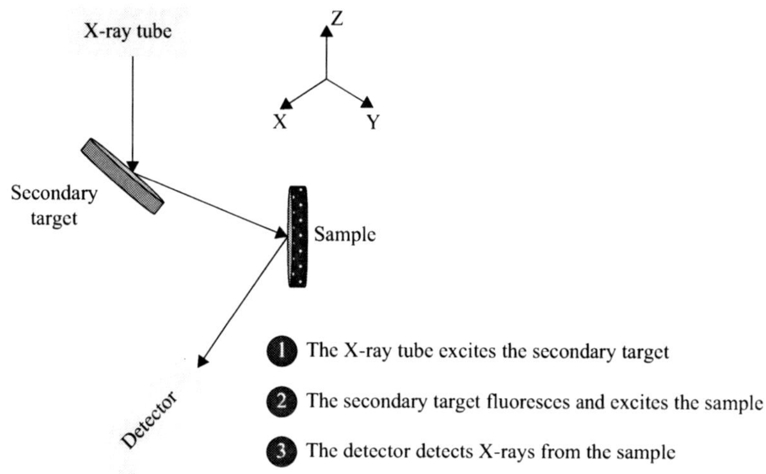

Figure 5.4 Geometric arrangement of a 3D-EDXRF system.

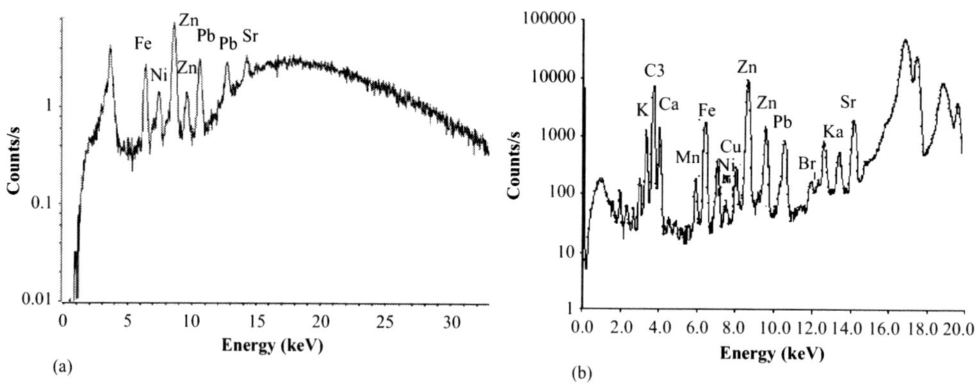

Figure 5.5. Basic spectra obtained in the analysis of a leaf sample using (a) 2D-EDXRF spectrometer (W X-ray tube) and (b) 3D-EDXRF spectrometer (W X-ray tube and Mo secondary target).

The commercialization of conventional EDXRF systems has increased in recent years, especially owing to instrumental technological advances, and at present a large number of benchtop and handheld EDXRF units (also called field-portable (FP)-EDXRF systems) are commercially available at a relatively modest cost investment.

Until recently, the use of radioactive X-ray sources (e.g., ^{55}Fe, ^{109}Cd, ^{241}Am) dominated in FP-XRF analyzers. These sources were well suited for use in portable devices as they are small (cylinders of approximately 5 mm height and diameter) and do not require external power. However, the short half-life of some isotopes, the strict regulatory requirements associated with transportation of radioactive materials, and their limited sensitivity available for trace metal determination represented significant barriers to their widespread adoption. Recent technological advances have, however, led to the availability of compact, low-power, metal-ceramic X-ray tubes that can be air cooled; these have largely replaced radioactive X-ray sources in low-cost handheld systems. Despite the fact that low-power X-ray tubes (50 W) present a lower sensitivity compared to high-power X-ray tubes (1–4 kW), the use of source modifiers (usually primary filters) and the new generation of compact-sized solid-state X-ray detectors (PIN diodes

Table 5.2. Analytical performance of different XRF instrumentation for the determination of metals preconcentrated on solid-phase extraction disks (Handheld—Tracer IV-GEO, Bruker AXS, Madison, USA; EDXRF—S2 Ranger Burker AXS, Germany; HE-P-EDXRF—Epsilon 5, PANalytical, The Netherlands)

	Linear range (μg/L)			Limit of detection (μg/L)		
Metals	**Handheld**	**EDXRF**	**HE-P-EDXRF[a]**	**Handheld**	**EDXRF**	**HE-P-EDXRF**
Ni	5–600	5–600	5–600	1.0	1.3	0.1
Cu	5–600	20–600	5–600	1.4	7.9	0.1
Zn	5–600	20–600	5–600	1.1	4.1	0.1
Pb	5–600	20–600	5–600	1.7	6.1	0.2
Cd	20–600	60–600	5–600	7.0	45	0.3[b] 0.6[c]

[a] HE-P-EDXRF, High-energy polarized-beam EDXRF.
[b] CsI secondary target.
[c] Al_2O_3 secondary target.

and SDDs) based on thermoelectrical cooling have improved the signal-to-noise ratio and the instrumental limits of detection and, at present, portable and handheld XRF units can offer results comparable to those of laboratory-based instruments. In Table 5.2, as an example, the analytical performance of different XRF instrumentation, including handheld, benchtop, and large-scale systems, for the determination of metals preconcentrated on solid-phase extraction disks is shown (Marguí et al., 2012).

As shown in Table 5.2, if ultratrace levels need to be determined in the preconcentrated aqueous sample of interest, the best option is to use a large-scaled instrument with high-power polarized X-ray tubes, especially if Cd determination is required. However, if metal concentrations are in the tenths or hundreds of microgram per liter range, then handheld or benchtop EDXRF systems, which offer extreme simplicity of operation in a low-cost compact design, are a good option. Moreover, with the use of handheld XRF units, it is possible to measure directly in the field (in-situ) and this can be particularly advantageous for acquiring near real-time data that is necessary to guide critical field decisions in the context of contamination, removal, and remediation actions. For this reason, these types of instrumentation have been employed as a common and standardized technique for on-site screening and fast turnaround analysis of contaminant elements in environmental samples, even if the elements are present at trace levels. Figure 5.6 shows an example of a spectrum recorded using a handheld XRF device (NITON®XLt™ 700 Series, Thermo-Scientific, Germany) obtained for the analysis of a soil contaminated with trace amounts of As (~20 mg/kg) (Parsons et al., in press).

Today, almost all manufacturers of XRF spectrometers worldwide provide EDXRF benchtop spectrometers and handheld units at a reasonable price. Even in the case of the more sophisticated EDXRF instruments (3D-optics) one can find benchtop prototypes (such as Phoenix II spectrometer by Spectro Analytical Instruments GmbH and the NEX CG by Rigaku Corporation). For certain, handheld and benchtop XRF units have given a new boost to the XRF technique. For more information about XRF manufacturers, readers should refer to the Buyers Guide in Chapter 9.

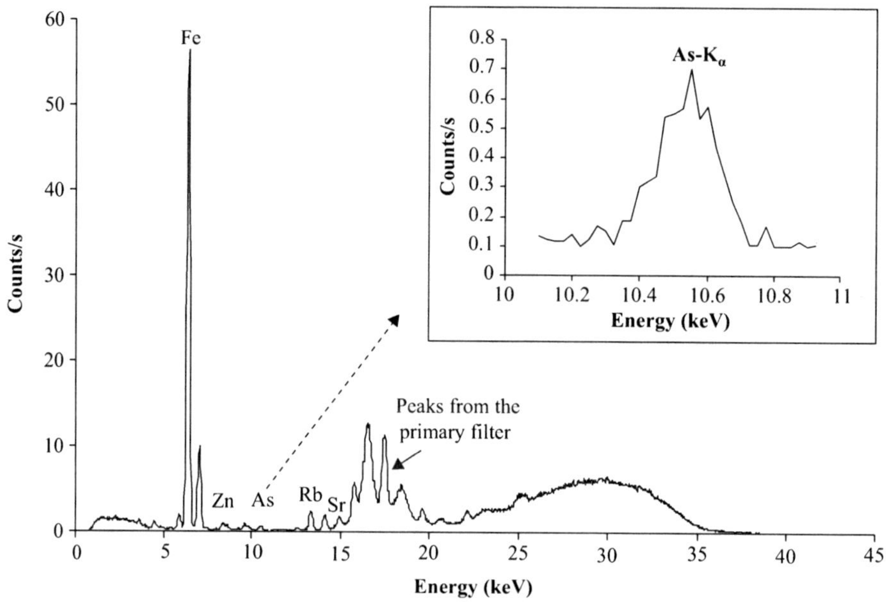

Figure 5.6. Example of a soil spectrum, with the As analysis range inset ([As] ~ 20 mg/kg), obtained using a handheld-XRF unit (NITON®XLt™ 700 Series, Thermo-Scientific, Germany).

Since the development of conventional EDXRF systems in the early 1970s, new instrumental variations of EDXRF spectrometers based on the design of novel X-ray optics have been designed. We will focus on this topic in Chapters 6 and 7.

5.3 COMPARISON OF WDXRF AND EDXRF SYSTEMS

The differences between WDXRF and EDXRF systems are because of not only the different detection systems but also other factors related to the instrumental layout. In this section, and in Table 5.3 as a summary, a comparison of the main instrumental characteristics for WDXRF and EDXRF spectrometers is shown.

Table 5.3. Main instrumental characteristics for WDXRF and EDXRF systems

Instrumental layout	**WDXRF**	**EDXRF**
Energy resolution	Good for light and middle atomic weight elements	Less optimal for light elements
Detecting efficiency	Low	High
Cost	Relatively expensive	Relatively inexpensive
Power consumption	50–1000 W	200–4000 W
Critical moving parts	Yes (crystal, goniometer)	No

- **Energy resolution:** Energy resolution in EDXRF systems is only related to the proportional characteristics of a photon detector used to separate the characteristic photons emerging from the sample and it is really poor over the light element range. (A possible exception is for "cryodetectors," which are cooled with liquid helium, namely *Superconducting Tunneling Junctions and Calorimeters*; these looked promising in the late 1990s but never made it commercially for a number of reasons, including their very high cost, low sensitivity, and the complexity of the spectra due to spurious peaks.) In contrast, WDXRF units employ diffraction by using a dispersing device to separate the sample's characteristic radiation. Therefore, using suitable analyzing crystals or a multilayer (with adequate *d*-spacing) dispersing elements, separation of close proximity lines is possible and the overlapping of light element lines is avoided.
- **Detection efficiency:** Owing to the simplicity of the EDXRF setup (only two basic units: excitation source and detection system) it is possible to position the detector closer to the sample and, therefore, the effective solid angle of the fluorescent X-rays is greater. In contrast, the complex configuration of WDXRF systems (equipped with collimator and dispersing devices) makes the solid angle, and thus the detecting efficiency, smaller compared to EDXRF units.
- **Cost:** Construction of EDXRF spectrometers is simpler compared to WDXRF systems as the complex optical system (collimators and dispersing device) is not necessary. Therefore, usually, EDXRF spectrometers are lower cost compared to WDXRF units.
- **Power consumption:** The brightness of a WDXRF spectrometer is very low in comparison to EDXRF units owing to the intensity losses in the analyzing crystal. To overcome this problem, the use of radiation sources of significant intensity (200–4000 W) is required. In contrast, the construction of low-power EDXRF systems (50 W) is possible.
- **Critical moving parts:** As we have already seen, in general the geometrical design of EDXRF spectrometers is much simpler compared to the WDXRF analogs, because, as stated earlier, they are composed of only two basic units: the excitation source and the detector and its associated electronics. In contrast, in WDXRF instruments there are some critical moving parts such as the analyzing crystal, the collimators, and the goniometer that lead to a less robust and flexible setup. This is the main reason why there is a greater range of benchtop, transportable, and handheld EDXRF systems compared to large-scale WDXRF spectrometers in the market.

Taking into account the different instrumental characteristics of WDXRF and EDXRF systems, it is easier to understand the analytical capabilities of these systems. In Table 5.4, an overview of the most significant analytical capabilities related to WDXRF and EDXRF units is shown.

- **Sensitivity and detection limits:** One of the problems of any X-ray spectrometer is that its absolute sensitivity decreases considerably as the lower atomic number region is approached (this is mostly due to the inherent low fluorescent yield of light elements). However, in WDXRF, owing to the use of suitable dispersion devices (with adequate *d*-spacing) and the extremely thin and high-transmission window of the gas flow proportional counter (detector), the sensitivity and detection limits for light elements are reasonable compared to those associated with conventional EDXRF systems.

Table 5.4. Main analytical capabilities for WDXRF and EDXRF systems

Analytical capability	WDXRF	EDXRF (2D)
Sensitivity and detection limits	Reasonable for light elements. Good for middle–heavy elements	Less optimal for light elements. Good for middle–heavy elements
Elemental range	Be to U	Na to U
Dynamic range	High	Moderate
Measurement	Sequential/simultaneous	Simultaneous

On the other hand, for middle–heavy elements, EDXRF methods are generally more sensitive with detection limits in the low milligram per kilogram range. However, when comparing the ED and WD type systems, the absolute sensitivity of the WDXRF is almost always higher than the equivalent value for an EDXRF system owing to the higher counting rates capability of the detector system and the possibility to select optimized conditions for element determination (analyzing crystal, measurement time, X-ray tube voltage and current, etc.).

- **Elemental range:** Taking into account the comments discussed above, usually when using WDXRF systems, the determination of light elements such as Be ($Z = 4$) is possible, while in conventional EDXRF spectrometers the lightest element that can be measured is often Na ($Z = 11$).
- **Dynamic range:** Generally, the dynamic range is higher in WDXRF systems mostly owing to the higher counting rates capability of the gas-filled or scintillation detectors used (up to 1–2 million cps) compared to the solid-state detectors associated with EDXRF systems (up to 50,000 cps). However, with the development of the new generation of Si drift detectors (SDDs), it is possible to record high count rates up to 300,000 cps without significant deterioration in the energy resolution.
- **Measurement:** In EDXRF, all of the detected radiation falls onto the detector at the same time and thus simultaneous multielemental information of the sample is recorded. We can also obtain multielemental information at the same time when using simultaneous WDXRF systems but the number of channels is limited and thus, as stated earlier, only the determination of 12–14 elements at the same time is possible.

Taking into consideration the comparative analytical capabilities for WDXRF and EDXRF systems, the selection of the most suitable configuration is based on the requirements for a given purpose. Some manufacturers, such as Oxford Instruments plc, have put into the market a series of hybrid element analyzers (e.g., MDX-1000) which contain the features of both EDXRF and WDXRF within a compact instrument. On the other hand, in recent years some new companies have appeared in the marketplace which offer not whole instruments as such but only some specialized instrumentation to design (in-house) instruments dedicated to special applications.

Both configurations (WDXRF and EDXRF) have been widely used for a large number of applications. In the following sections, some case studies of WDXRF and EDXRF applications in the industrial and environmental fields are presented.

5.4 APPLICATIONS OF WDXRF AND CASE STUDIES

Multichannel WD instruments are used almost exclusively for routine, high-throughput analyses when there is a need for fast accurate analysis, but where flexibility is of little importance. An example of this is the wide use of such systems to determine the chemical composition of all materials involved in cement production or in the metallurgy industry. The potentiality of WDXRF for quality control analysis of several elements in commercial milk-based products and pharmaceutical samples has also been demonstrated in recent years.

In spite of the predominant use of more expensive WDXRF systems in routine industrial analysis, the use of such systems in the environmental field is also interesting. For instance, WDXRF is an extremely popular tool in the geological investigation field and it has been used to measure raw materials and finished products composed of minerals, soils, sediments, and wastes. Other studies highlight the possibilities of using the technique for the quantitative analysis of macro and micronutrients in vegetation matrices as an effective alternative to atomic spectroscopy, which requires a previous digestion of the vegetation sample before analysis. Furthermore, the possibility of determining some elements that can hardly be determined by other spectroscopic techniques (such as S) is also and additional advantage. Analytical procedures using WDXRF for trace elemental determination in environmental water have also been published; usually, an additional preconcentration procedure is needed before the WDXRF analysis.

Below, some examples of WDXRF applications in different fields are presented and some guidelines for laboratory work are discussed.

5.4.1 DETERMINATION OF METAL RESIDUES IN ACTIVE PHARMACEUTICAL INGREDIENTS

At the beginning of 2008, the European Agency for the Evaluation of Medicines (EMEA) presented a guideline (EMEA/CHMP/SWP/4446/2000) on the specification limits for residues of metal catalysts or metal reagents in active pharmaceutical ingredients (APIs). Sample preparation is usually a critical step in pharmaceutical samples because of the poor solubility of the matrices. A promising alternative is the use of a solid-state analysis technique such as WDXRF. In such a case, sample preparation includes a first step of homogenization and then the pressing of the powdered material (1 g) into pellets without any chemical treatment. Therefore, by comparison, it involves less manipulation and fewer time-consuming digestion steps, implying cost and time savings for the manufacturers. Additional advantages of this simple sample preparation procedure are the possibility of reanalysis and reuse of the pelletized ground material if necessary.

For quantification purposes, the use of several synthetic standards made of cellulose to simulate an API matrix is an effective means to obtain reliable calibration curves when no suitable pharmaceutical reference materials are available. The calculated limits of quantification (LOQ) (low milligram per kilogram range) for Fe, Zn, Cr, and Ni comply with the rigorous limit requirements of current legislation (Marguí et al., 2009b). Moreover, it has been demonstrated that the proposed methodology conforms to the International Conference on Harmonisation (ICH) of Technical Requirements for Registration of Pharmaceuticals for Human Use validation guidelines[1] required to test any analytical procedure used for the analysis of impurities in pharmaceuticals for human use. All these capabilities make this analytical technique very attractive for implementation within the pharmaceutical industry in order to increase the productivity of a laboratory.

5.4.2 DETERMINATION OF HEAVY METAL CONTENT IN AUTOMOTIVE SHREDDER RESIDUES (ASR)

Every year between 8 and 9 million vehicles in the European Union arrive at the end of their useful life. Car wastes can have very high metal content, thereby falling into the hazardous waste class. Consequently, the characterization of such residues in terms of their chemical composition is important. One of the main drawbacks of ASR wastes is their large heterogeneity, as they are composed of metallic wires, paper, sand, wood, and also large amounts of rubber and foams.

An analytical option for determining the heavy metal content in ASRs is the use of a WDXRF method. Samples (previously sieved to the required grain-size) are mixed with a binding agent (methyl methacrylate acrylic resin, 0.4 g) and then pressed to obtain pellets for further analysis. The metal content evaluation from WDXRF spectral data can be carried out by using the fundamental parameter method. This approach is suitable for the analysis of materials when done does not have appropriate certified reference materials to calibrate an analytical laboratory procedure, thus enabling accurate standardless analysis (Gonzalez-Fernandez et al., 2008).

5.4.3 METAL DETERMINATION IN POLLUTED SOILS AND WATERS

One of the critical factors for successfully conducting contamination characterization, removal, and remedial operations at metal-contaminated sites by mining operations is the potential to analyze samples in a timely and appropriate manner. Rapid, multielement analysis for minor and major elements can be performed using XRF spectrometry. An additional advantage is the possibility to perform the analysis directly on the solid samples, which is really convenient when dealing with environmental samples that are difficult to dissolve such as soils, sediments, and mining wastes (Gonzalez-Fernandez, Jurado-Roldan, and Queralt, 2011). The sample preparation for this kind of sample involves the preparation of a fused bead or a pressed pellet of the powdered material for the determination of major light elements or minor and trace elements, respectively. The calibration curve method employing a set of standards with matrices similar to the target samples is perhaps the best option for quantification. At present, there is a large variety of geological standards listed in published catalogs that can be used for this purpose.

In contrast, the use of a WDXRF system can also be useful for the analysis of liquid samples involved in mining environments. When a mine is abandoned, the water pumping ceases and the water floods the mine. The result is water containing elevated levels of potentially toxic metals. Hence, the monitoring of elements in these kinds of liquid samples is environmentally significant; this is a difficult task when using conventional atomic spectroscopic techniques owing to the complex matrices of these waters. Sample preparation for further WDXRF

[1] http://www.ich.org/fileadmin/Public_Web_Site/ICH_Products/Consideration_documents/GTDG_Considerations_Documents/ICH_Considerations_General_Principles_Risk_of_IGI_GT_Vectors.pdf

analysis is quite simple and consists of deposition of 1–6 mL of a water sample in special liquid sample holder that incorporates a tapered snap-on ring at the end of the cell for attachment of a thin-film liquid sample support (4.0 µm thick Prolene™ X-ray foil). The specimen are then dried using an infrared lamp (to speed up the sample drying time) and sealed in the sample holder of the spectrometer for WDXRF analysis. For quantification purposes, aqueous standards containing different amounts of elements can be prepared using the same treatment procedure. Another option is to use internal standardization and thus, no external calibration is required and samples can be quickly screened without any restrictions due to matrix effects and calibration ranges (Gonzalez-Fernandez, Marguí, and Queralt, 2009). With this methodology, the calculated detection limits for most middle and heavy elements are in the low milligram per liter level (depending on the element) that is adequate for element determination in most metal mining liquid effluents.

5.5 APPLICATIONS OF EDXRF AND CASE STUDIES

ED spectrometers have the great advantage of being able to display information on all elements at the same time. They lack somewhat in resolution over the WD spectrometer, but the ability to reveal which elements are absent and which ones present make an ED spectrometer ideal for general troubleshooting problems. They have been particularly effective, inter alia, in the fields of scrap alloy sorting, forensic science, and the provision of elemental data to supplement X-ray powder diffraction data. The application of portable EDXRF has also found special use in art, archaeology, and numismatics studies owing to the possibility of analyzing objects in situ that cannot be brought into the laboratory.

An area where these instruments are finding increasing application is in the analysis of pollutants. More recent developments have achieved sensitivities that are quite sufficient for the determination of trace metals in environmental samples. Good examples of this are the analysis of the chemical composition of atmospheric aerosols, vegetation samples, environmental waters (after a preconcentration procedure of the sample), and soils. In this latter case, the US-EPA (Unites States Environmental Protection Agency) also issues a Standard Operation Procedure (SOP) for elemental analysis of metals in soils and sediments using a portable EDXRF analyzer.[2]

More specific EDXRF systems, such as micro-EDXRF spectrometers, also have wide specific applications; they allow one to obtain analytical information at the microscale level, which is of particular significance for studying the elemental distribution in heterogeneous environmental samples. We will discuss some applications in Chapter 7. In the following sections, some applications of conventional EDXRF and 3D-EDXRF systems are presented.

5.5.1 DETERMINATION OF HEAVY METALS AT TRACE LEVELS IN VEGETATION SAMPLES

Determination of the chemical composition of vegetation matrices in the environmental and industrial field has become increasingly more important over the last few years. Analysis by

[2] http://www.epa.gov/osw/hazard/testmethods/sw846/pdfs/6200.pdf

XRF spectrometry complies with the desired features for the analysis of vegetation specimen including the need of multielement capability, a wide dynamic range, high throughput, and low cost per determination. Moreover, sample preparation of vegetation specimen is quite simple: crushing and/or grinding of the plant material (previously oven-dried at $T \leq 60°C$) into fine powder followed by a pelletization at high pressure.

However, one of the main drawbacks of conventional EDXRF instrumentation, restricting a more frequent use, has been the insufficient sensitivity for some important pollutant elements, such as Cd. This is related to the high background intensities attributable to the high degree of scattering of the X-ray source by organic matrices and the problems of reduced sensitivity and spectral interferences inherent to the choice of L_α lines for the determination of heavy metals such as Cd. An interesting approach to overcome this problem is the use of polarized-beam EDXRF, as mentioned earlier. The use of suitable secondary targets for excitation not only allows for reducing the intensity of the measured continuum radiation, thus significantly increasing the signal-to-noise ratio, but also to increases X-ray production owing to the choice of quasi-monochromatic radiation with energy close to the photoelectric absorption edge of a given element of interest. Therefore, compared to other EDXRF techniques, it has the advantages of increased sensitivity (net intensity per unit of analyte concentration) and specificity for trace element analysis. Moreover, with novel instrumentation the use of a Gd anode X-ray tube and a high-energy Ge semiconductor detector allows for undertaking EDXRF analysis using K-lines of high-Z elements such as Cd. Using such instrumentation, it was found that using the best analytical conditions (Al_2O_3 as secondary target + Mo primary filter, 100 kV, 5 mA, 1000 s measuring time), a limit of detection for Cd of 0.6 mg/kg was achieved (Marguí et al., 2006). Other trace elements (As, Pb, Fe, Cu, and Zn) were also determined using optimized analytical conditions and using a standardless fundamental parameter approach. The accuracy and precision in all cases were acceptable for this kind of study.

5.5.2 DETERMINATION OF Cu, Ni, Zn, Pb, AND Cd IN AQUEOUS SAMPLES

The use of solid-phase extraction disks for metal (Cu, Ni, Zn, Pb, and Cd) preconcentration in aqueous samples prior to analysis by EDXRF spectrometry is a good analytical approach to improve the limits of detection compared to the method of direct analysis of liquid samples. While the extraction disks are ideal supports (thin-layer organic materials) for sample analyses by XRF, the collection of the elements can be performed by means of a simple filtration procedure (using 1 L of aqueous sample) which permits the possibility of performing the direct measurement in the field (in-situ) if a portable (handheld) EDXRF instrument is used.

This procedure has been successfully applied, in combination with different XRF systems, for the analysis of analytes in tap and waste water samples. If ultratrace levels need to be determined in the aqueous sample of interest, the best option is to use a large-scaled laboratory-based instrument with high-power polarized X-ray tubes, especially if Cd determination is required. However, if metal concentrations are in the tenths or hundreds of micrograms per liter range, then a handheld or benchtop EDXRF system, which offers extreme simplicity of operation at a low-cost compact design, are a good option. Moreover, with the use of handheld XRF units, the direct measurement in the field (in-situ) is possible and that could be particularly important for providing near real-time data necessary to guide critical field decisions concerning contamination, removal, and remediation actions (Marguí et al., 2012).

5.5.3 *CHEMICAL CHARACTERIZATION OF AEROSOL SAMPLES*

XRF is frequently used for trace analysis in air pollution studies. In the air, trace elements are almost exclusively in the particulate phase at typical concentrations of 50–500 ng/m^3. By simply drawing a large air volume through a filter, large preconcentration factors are easily achieved. Usually Teflon® or polycarbonate (Nuclepore®) filters are used because of their high purity and because they are surface collectors. Then, adequately loaded filters with air particulate matter on them are presented directly to the EDXRF unit. Unless the filters are extremely loaded, no correction for the X-ray absorption in the aerosol material is usually necessary. However, for light elements such as Si, P, S, and Cl, the absorption effect needs to be corrected for by using a suitable correction procedure. Aerosols analysis has been a very important application for EDXRF since the 1970s.

As industrially based societies have become more aware of the extent of particulate matter that we inhale, a reduction of the regulatory limits has followed, thereby challenging analysts to develop methods with sufficient sensitivity to meet any new need. In this respect, in addition to conventional EDXRF, energy dispersive polarized-beam EDXRF spectrometers have also been applied to aerosol analysis. As previously stated, by combining polarization with the possibility of different secondary sources, an instrument can be designed to generate a range of excitation and polarization conditions optimized for different groups of analytes. For instance, a methodology for the quantification of 20 elements deposited on different aerosol filters using such instrumentation was published in 2005 (Spolnik et al., 2005).

CHAPTER 6

Total Reflection X-Ray Spectrometry (TXRF)

6.1 INTRODUCTION AND BASIC PRINCIPLES

Since its development, one of the drawbacks of conventional EDXRF that has hampered the determination of ultratrace element levels in small samples has been the high background due to scatter from a sample. In 1971, an analytical note published in *Review of Scientific Instruments* by Yoneda and Horiuchi (Yoneda and Horiuchi, 1971) reported the benefits of the total reflection of X-rays at an optically flat surface for spectrochemical analysis. The basic principle was that if a low, divergent, almost parallel X-ray beam impinges on the flat, smooth surface of a reflector at an angle smaller than the critical angle ($\varphi_{crit} \sim 0.1°$), total reflection occurs. Under such conditions, most of the primary beam radiation is reflected and thus the spectral background is clearly reduced. This effect is graphically shown in Figure 6.1.

In addition to the significant reduction of the spectral background, the fluorescence signal is also enlarged as the reflected beam contributes to excitation, leading to a doubling in the intensity of the fluorescence signal. For these reasons, detection limits of total reflection XRF (TXRF) systems are usually at least one order or magnitude lower than those associated with conventional EDXRF.

TXRF is thus a variation of EDXRF in which the primary beam strikes the sample at a glancing angle of less than 0.1° and not at the typical EDXRF angle of about 40°. Moreover, owing to the grazing incident excitation angle, the detector can be positioned very close to the sample leading to a large solid angle for the detection of the fluorescence signal (see Figure 6.2). This also contributes to an improvement in the sensitivity of TXRF systems compared to conventional EDXRF spectrometers.

Taking into account the basic principles and the analytical capabilities of TXRF, since its development it has been primarily used for chemical micro and trace analysis of small quantities of solutions and suspensions placed on optical flats as sample supports. However, owing to the angle dependence of the fluorescence signal in the range of total reflection, since the 1980s

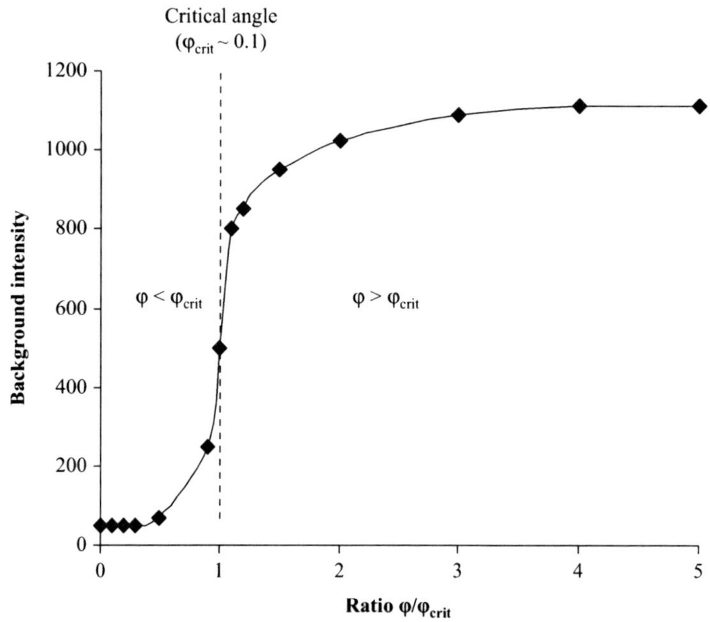

Figure 6.1. Effect of incident angle on background intensity. φ, incident angle; φ_{crit}, critical angle.

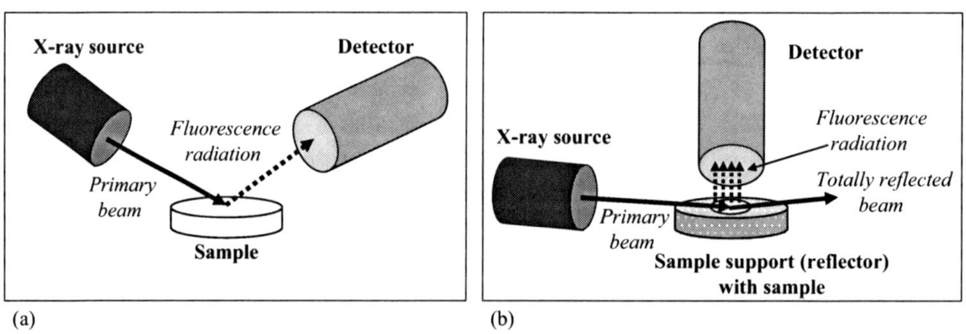

Figure 6.2. Instrumental arrangements for (a) conventional EDXRF system and (b) TXRF system.

TXRF has also been used to investigate surface impurities and even molecules adsorbed onto flat surfaces. Such examinations were and still are of relevant interest in the semiconductor industry to study possible metal impurities on silicon wafers.

In this chapter, the instrumental aspects of TXRF and analytical capabilities for chemical and surface analysis using the technique will be discussed. Moreover, at the end of this chapter, a summary of the main applications of TXRF and some case studies are presented. Readers interested in more detailed information about the TXRF technique should refer to an earlier book that focuses on this topic (Klockenkämper, 1997). Very good overviews on recent developments and applications in the TXRF field can also be found in some published journal reviews (Streli, 2006; Von Bohlen, 2004).

6.2 TXRF LAYOUT

In Chapter 2, the main types of sources, modifiers, and detector systems were presented in detail. In this section, only the spatial configuration of a TXRF system will be considered. Therefore, for more detailed information on the basic components of TXRF units, readers should refer back to Chapter 2.

There are several designs of TXRF spectrometers, but the geometric arrangement of the components in a TXRF instrument for general laboratory use is shown in Figure 6.3.

The usual excitation source for TXRF is an X-ray tube equipped with a Mo anode, but for some applications targets of Cu (determination of light elements), Ag, and W (determination of high-Z elements) are also used. The polychromatic collimated beam of the X-ray tube is reflected to a source modifier to suppress the high-energy photons (above the K lines of the anode material), whereas the low-energy part of the primary radiation, including the characteristic X-ray lines, are reflected, improving the peak-to-background ratio in the low-energy region. For most applications, a flat, polished quartz block (named *cut-off reflector*) is satisfactory for such a purpose. Alternatively, this first source modifier can be substituted by a monochromator in which the undesirable parts of the continuous spectrum are eliminated according to Bragg's law (see Section 2.5.4 in Chapter 2). Only the beam reflected at this spectrum modifier is allowed to impinge onto the sample carrier at a rather small glancing angle (below the critical angle) to ensure the condition for total reflection. The critical angle can be calculated as follows:

$$\phi_{\text{crit}} \approx \frac{1.65}{E}\sqrt{\frac{Z}{A}\rho} \qquad (6.1)$$

where E represents photon energy (kV), Z is the atomic number of the reflector, A is the atomic mass (g/mol), and ρ is the density (g/cm^3). As can be deduced from Equation 6.1, the values of the critical angle depend on the incident energy of the photons and the composition of the reflector on which the sample is deposited. This effect is clearly shown in Table 6.1, where critical angle values for various types of sample carriers and different photon energies of the beam are listed.

Usually, the sample holder is loaded with some sample material (in Section 6.3.1.3 the most common sample treatment procedures to present samples for TXRF analysis are discussed) or a substrate representing the actual object of analysis itself is presented. In either case, to ensure

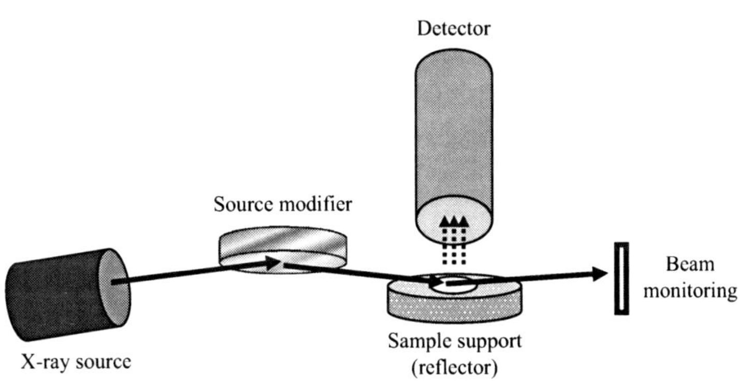

Figure 6.3. Geometric arrangement of a TXRF system.

Table 6.1. Critical angles (α_{crit}) in degrees calculated for various media and X-rays of different photon energies (modified from Klockenkämper (1997)).

Medium	8.4 keV	17.44 keV	35 keV
Plexiglas®	0.157	0.076	0.038
Glassy carbon	0.165	0.080	0.040
Quartz	0.21	0.10	0.050
Silicon	0.21	0.10	0.051

that the analysis is performed under total reflection conditions both the source modifier and the sample need to be precisely positioned. Therefore, an optimal adjustment of the TXRF setup is needed to ensure good analytical results. Usually, the sample carrier is mounted on a plate holder that may be tilted so as to reach a parallel plane and later on to adjust its tilt with respect to the beam line. Three screws in combination with springs can be used for this purpose. The interaction of the beam with the sample can be monitored by observing the resulting shape of the beam and the appearance of the reflected beam using a ZnS fluorescent screen or a charge-coupled device (CCD) camera that allows the conversion of optical brightness into electrical amplitude signals.

For some applications (surface and near-surface layer analysis, for more detail see Section 6.3.2), a special TXRF setup allowing measurements at various angles is required. These systems are usually equipped with monochromators (instead of a cut-off reflector) to produce an unambiguous dependence of the intensity and the glancing angle (for more information on monochromators see Section 2.5.4). Usually, a wafer robot is used for loading and changing the sample wafers (∅ = 100–200 mm). Scan measurements are performed starting from a setting close to zero and increasing the incident angle stepwise.

Finally, in all cases, the fluorescence intensity arising from the sample is recorded by an ED solid-state detector, usually a Si(Li) detector. As shown in Figure 6.3, the detector is mounted perpendicularly to the sample plane to obtain a decreased background in spectra, and it is also placed very close to the sample (about 1 mm distance) to enlarge the solid angle (as already mentioned in Section 6.1). Then the measured signal, as in all EDXRF systems, is sorted by amplitude in a multichannel analyzer, leading to an ED spectrum. In some TXRF designs, to reduce attenuation of the low-energy characteristic emission radiation in air, measurements can be performed under vacuum or He flush (using appropriate sample chambers). In Figure 6.4, as examples, basic spectra obtained in the analysis of environmental water samples using TXRF instrumentation equipped with W and Mo X-ray tubes are displayed. As is shown, for both TXRF configurations, the spectral background is clearly reduced compared to that associated with a conventional EDXRF configuration (see Figure 5.5 in Chapter 5).

After the description of the first setup for TXRF by Yoneda and Horiuchi, the technique was further pursued by Aiginger and Wobrauschek, in the early 1970s (Aiginger and Wobrauschek, 1974). The first commercially available TXRF spectrometer was designed in 1982/1983 and since then instruments have been improved continuously in parallel with the technological improvements related to excitation and detection devices.

Over the last decades, most of the published TXRF analyses were performed using large-scale instruments with high-power X-ray tubes that required water-cooling systems and

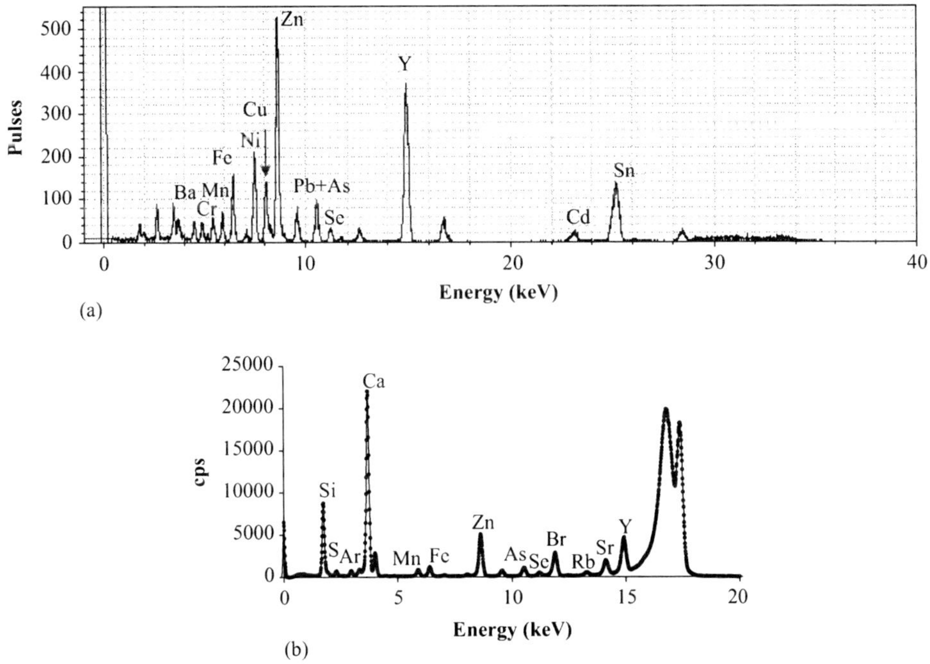

Figure 6.4. Basic spectra obtained in the analysis of environmental water samples using TXRF instrumentation: (a) W-anode X-ray tube, 50 W, 1000 s, 10 μL of sample, quartz reflector (S2 PICOFOX spectrometer, Bruker AXS), (b) Mo-anode X-ray tube, 3 kW, 1000 s, 10 μL of sample, quartz reflector (TXRF 8030C spectrometer, Atomika Instruments).

liquid-nitrogen cooled detectors. However, in recent years, the development and commercialization of benchtop TXRF instrumentation, which offer extreme simplicity of operation in a low-cost compact design, have promoted its application in industry as well as in research activities.

So far, we have focused on TXRF systems using X-ray tubes as excitation sources. However, instead of conventional X-ray tubes, synchrotron radiation may be used for excitation. When using this brilliant photon beam, the detection power is significantly improved (femtogram scale) compared to commercially available instruments (detection of nanograms down to a few picograms). But the respective large scale, usually national facilities, storage rings for electrons and positrons necessary for this purpose are restricted and this kind of beam line is usually employed only for research purposes. In Chapter 7, a more detailed description of synchrotron radiation-induced X-ray emission is displayed.

6.3 ANALYTICAL CAPABILITIES OF TXRF SYSTEMS

Considering the instrumental characteristics of TXRF, as discussed earlier, it is easier to understand the analytical capabilities and applications of such systems. The most significant capabilities are:

- **Multielemental capability:** Similarly to conventional EDXRF systems, in TXRF units the fluorescence radiation emerging from the sample falls onto the detector at the same time and thus simultaneous multielemental information of the sample is obtained.

- **Microanalytical capability:** Specimen need to be deposited on totally reflecting carriers in only small amounts (few microliters or micrograms). Therefore, the TXRF technique can be used as a microanalytical tool for mass-limited samples.
- **Limits of detection:** Owing to the improvement in signal-to-noise ratio, the instrumental detection limits for most commercial spectrometers are typically in the range of picograms or nanograms per milliliter. These amounts are more than two orders of magnitude lower than those detectable by conventional XRF methods. Absolute limits of detection can be improved to a few femtograms using TXRF with excitation by means of synchrotron radiation.
- **Easy quantification:** As the sample constitutes a very thin layer, the quantification process is less prone to matrix influence and the correction for attenuation or enhancement effects, commonly required in EDXRF and WDXRF analysis, is not necessary (for more details see Section 6.3.1.3).
- **Angle dependence of fluorescence signal:** By measuring the angle dependence of the fluorescence signal, the location of the fluorescing atoms above or below the surface is possible. Therefore, TXRF allows for discriminating between a residue on a surface, a surface layer, and a bulk sample. In such experiments, the sample must be tilted in several steps to record angle-related scans.

In addition to all these analytical properties, commercial TXRF systems are rather simple, versatile, and cost effective. The final purchase cost depends mostly on the configuration of the TXRF setup (one angle position or angle scan available) and also on the instrument-size and the maximum power of the X-ray tube. Currently, benchtop systems are equipped with compact size solid-state X-ray detectors (PIN diodes and silicon drift detectors) based on thermoelectrical cooling and with low-power metal ceramic X-ray tubes, which can be cooled by air. Therefore, water and liquid-nitrogen cooling systems are not needed and these kinds of spectrometers can be purchased at a very modest cost investment.

The convenient multielement microanalytical capability of TXRF has promoted its use for surface analysis, microanalysis, or chemical analysis purposes in many fields. In the following subsections, the main possibilities and drawbacks of TXRF for these applications are presented. In Section 6.4, some application cases in different fields will be discussed in more detail.

6.3.1 CHEMICAL ANALYSIS

For chemical analysis applications, TXRF has been mostly applied for the analysis of small amounts of liquids (with or without a preconcentration treatment) and solid samples (after a decomposition, dissolution, or suspension procedure) deposited onto a carrier with high reflectivity that serves as a reflecting sample support. Detailed information on types of sample carriers, sample treatment procedures for the analysis of liquid and solid samples, and quantification approaches in chemical analysis by TXRF are presented here.

6.3.1.1 Sample Carriers

As already stated in Section 6.2, the values of the incident angle necessary to work under conditions of total reflection depend on the composition of the reflector used to deposit the microliters of liquid sample or suspension for TXRF analysis. From the chemical point of view, the

Table 6.2. Advantages and disadvantages of different types of sample carriers

Medium	**Advantages**	**Disadvantages**
Plexiglas®	Inexpensive Low blank values	Very high background Disposable Frequently S blank value
Quartz glass	Very hard Very pure Easy to clean	Not stable against HF Si blank value Quite expensive purchase
Sapphire	Chemically very stable Durable	Al blank value Expensive

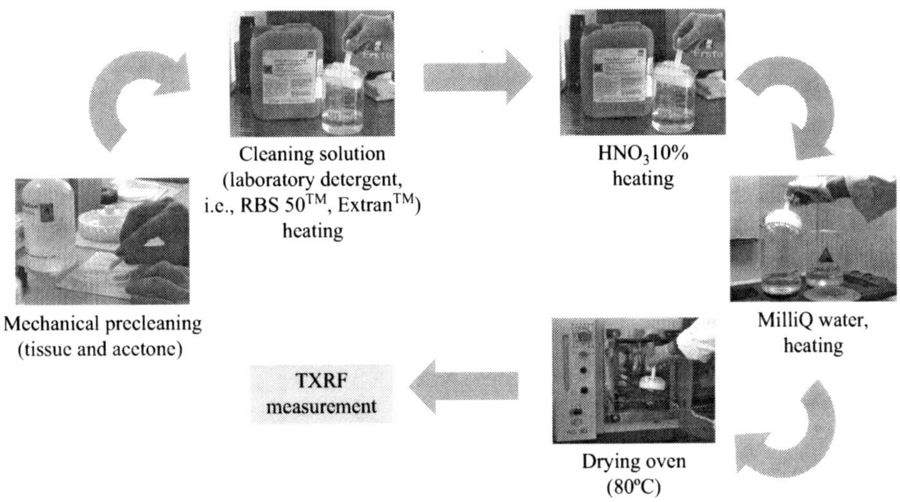

Figure 6.5. Main steps followed for the cleaning of sample carriers (Pictures from S2 PICOFOX, Bruker AXS).

selection of the type of sample carrier depends on the elements to be measured, the chemical resistance of the reflector material, and on the spectral background associated with the reflector material. In Table 6.2, the advantages and disadvantages of the most commonly used types of sample carriers are listed. Usually, reflectors used as sample holders for introducing samples into TXRF equipment are disks with a diameter of 30 mm and a thickness of 3 mm ± 0.1 mm.

As the sample carriers play an important role for the achievement of optimal analytical results, the material and also its mechanical performance are crucial. Despite the fact that the reflectors are usually made of very hard material, they can still be damaged by inappropriate handling. This can alter their reflectivity leading to an increase in the spectral background.

When using nondisposable reflectors (e.g., quartz), a proper cleaning of its surface has to be ensured to avoid memory effects, particularly if trace elements have to be determined. Contaminations on a sample carrier will automatically result in errors in both qualitative and quantitative element determination. In the course of time, many TXRF users have developed individual procedures for the cleaning reflectors. In Figure 6.5, a view of the main steps commonly followed is shown.

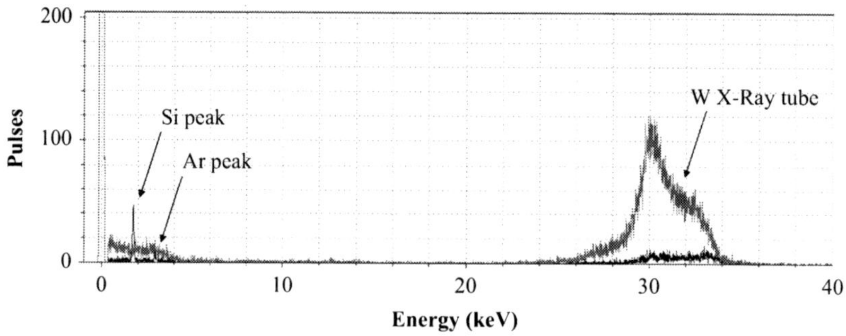

Figure 6.6. Typical spectra of a clean quartz glass sample support (•) and of a clean Plexiglas® reflector (•) in analytical conditions of W X-ray tube, 50 kV, 1 mA, and 1000 s).

After the drying step (see Figure 6.5), nonhydrophobic surfaces (e.g., quartz reflectors) must be coated with a silicone solution and subsequently dried at 100°C to favor the deposition of the target sample within a small spot on the cleaned reflector.

The cleaning time for each step has to be tested but, usually, to ensure a thorough complete cleaning of sample carriers, the whole procedure takes about 2–3 h.

After the cleaning procedure, a blank spectrum should be measured for each reflector to ensure the absence of any contamination from the previously analyzed sample. In Figure 6.6, typical spectra of a clean quartz reflector (black) and of a clean Plexiglas® support are displayed. Only system inherent signals are detected: Si from the quartz reflector and, in both cases, Ar from the air gap between the reflector and the detector, and the scattered primary radiation of the W X-ray tube.

6.3.1.2 Sample Treatment Procedures for Chemical Analysis by TXRF

As mentioned already, TXRF has been mostly applied for the analysis of small amounts of liquid and solid samples. Since the development of TXRF, several treatment procedures have been proposed for such purposes. However, in all cases, the aim of the preparation process is to obtain the test specimen as a thin layer (<100 μm) on a carrier with high reflectivity. Therefore, the choice of appropriate sample preparation is of crucial importance for TXRF analysis to obtain a thin layer and to ensure the conditions of total reflection. In Figure 6.7, a sample preparation flow chart for liquid and solid materials is presented.

- **Liquid samples**

For low salinity liquid samples, the TXRF analysis can be directly performed by depositing a small volume of the liquid sample onto a suitable reflector and drying. If quantification by internal standardization is performed (for more details about internal standardization, see Section 6.3.1.3), then an aliquot of 500–1000 μL of the original liquid sample is transferred to a container with a pipette, and a suitable amount of an internal standard solution is then added. Afterward, the mixture is thoroughly homogenized (i.e., in an automatic sample shaker) before the deposition and drying of the sample onto the reflector. If liquid samples are only available in

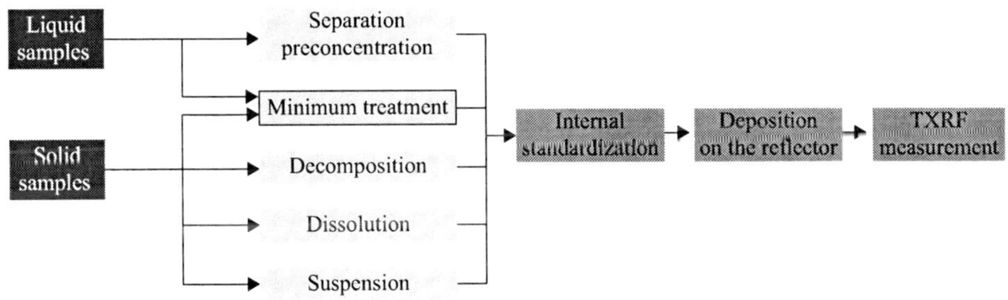

Figure 6.7. Sample preparation procedures for the analysis of liquids and solids by TXRF.

such small amounts that an internal standardization/homogenization in a sample container is not possible, the internal standard may be added onto the sample carrier subsequently. However, in such cases, a homogeneous distribution of the internal standard cannot be guaranteed. It is also important to note that for the analysis of organic liquid samples, the added internal standard has to be based on an organic solvent in order to obtain a homogeneous distribution of the internal standard within the organic sample.

Usually, the diameter of the sample spot on the sample carrier has to be within the X-ray beam size (≤10 mm) to ensure completely exposing of the drop to the X-ray beam. For that, only a few microliters (5–20 μL) of solution is required. To increase the sensitivity for a given sample, in some cases, multiple droplets are deposited sequentially at the same location with time allowed for each droplet to dry between successive depositions. The drying of the sample deposited onto the reflector is usually performed under vacuum in a desiccator or by heat with IR-lamps, laboratory ovens, or heating plates. Recent advances covering the topic of sample deposition are focused mainly on the use of nanodroplets or picodroplets for dried spot preparation to attain a more reproducible spot deposition with decreased drying times (20–170 s for 2 nL droplets) (Miller and Havrilla, 2004). Moreover, the resulting drying spots are uniform in shape and elements contained within them can be detected at picogram detection limits. This new sample preparation method has the potential for use in situations where sample availability is limited.

However, when dealing with the analysis of more complex samples (e.g., samples with high salt content) additional sample treatments have to be used to obtain an adequate presentation of the sample to comply with the thin layer condition. Depending on the concentration range, dilution of a liquid sample may be helpful to ensure a homogeneous thin layer of the complex sample on the reflector. Usually, dilutions using ultrapure water or diluted solutions of a commercial nonionic detergent (i.e., Triton® X-114) are used. The dilution factor depends mostly on the type of sample and also on the concentration range of the elements to be measured. For instance, for blood sample analysis, 500 μL of a sample is usually mixed and homogenized with 500 μL of ultrapure water. Then, further preparation has to be executed in the same way as discussed earlier. If determinations of extremely low-element concentrations have to be undertaken, then the use of preconcentration procedures is required before the TXRF analysis is performed. In addition, to improve limits of detection for element determination, when using more sophisticated sample treatments before a TXRF analysis, the determination of volatile elements and the attainment of element speciation information is also desirable. However, it is important to remark that when using additional sample treatments, in most cases, obtaining multielemental information is no longer possible (most of the treatments are selective for one or a small group of elements), a higher sample volume is needed, and it is also especially necessary to guard against losses and contamination by other elements that can lead to systematic errors on the final results.

A large number of sample treatment procedures have been developed and reported to preconcentrate analytes prior to TXRF measurements. Most of them have been already used in combination with other atomic spectroscopic techniques that also involve the analysis of liquid samples (e.g., flame atomic absorption spectroscopy (FAAS) or inductively coupled plasma atomic emission spectroscopy (ICP-AES)). A comprehensive overview on the topic can be found in an article published some years ago (Prange and Schwenke, 1989). But perhaps the most commonly used preconcentration procedures in combination with TXRF spectrometry are based on complexation and extraction of the analytes. In Figure 6.8, the main steps involved in these analytical procedures are shown.

As is shown in Figure 6.8, most procedures using solid–liquid extraction and liquid–liquid extraction for subsequent TXRF analysis are based on the complexation of elements under investigation using carbamates, followed by solvent extraction with an organic reagent. Then, the water-immiscible liquid phase is collected and dried onto a sample carrier before analysis.

In recent years, some new efficient variations of these methods and new techniques extending the possibilities of TXRF for trace element analysis in liquid samples have been published. However, there is no panacea, no ideal and universal method of sample treatment applicable over an infinite concentration range capable of determining all elements but which is also simple, fast, and cost-effective. Therefore, the final choice will be influenced by the types of liquid

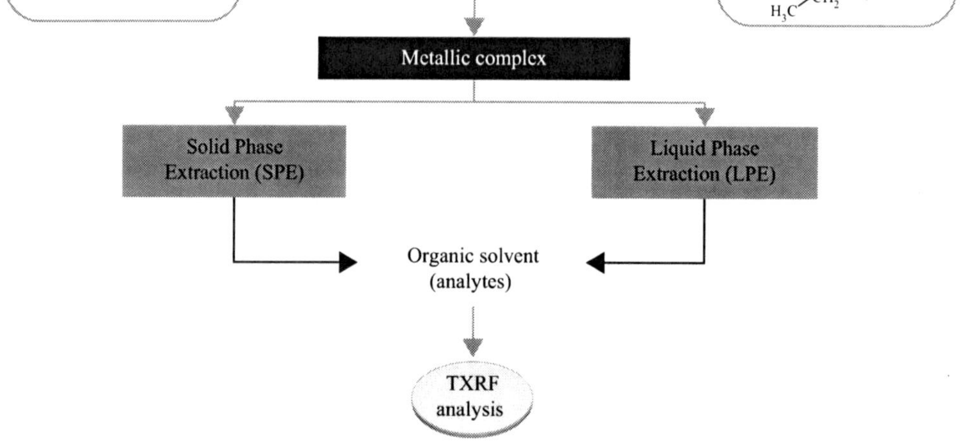

Figure 6.8. Main steps of preconcentration procedures based on complexation and extraction of the analytes prior to TXRF analysis.

that have to be examined, the elements of interest, the number of samples, the required accuracy, precision, and sensitivity, and so on.

The research group led by Kallithrakas-Kontos in Greece developed selective membranes containing a few micrograms of various complexing reagents within a polymeric matrix to be used as a preconcentration tool prior to a TXRF analysis. The membranes were placed on the center of quartz reflectors and they were used to collect trace elements from various water solutions of volumes from 5 to 500 mL. The cleaning procedure after TXRF analysis was simple and the membranes were easily removed from the quartz surface. Additional advantages of these membranes included the small quantities of the reagents required to create the membranes (micrograms or microliters). This methodology has been applied by the same research group for the selective or multielement analysis of metals in different types of water samples (including seawater) with detection limits in the low microgram per liter range, even for volatile elements such as Hg (Koulouridakis and Kallithrakas-Kontos, 2004).

Another interesting approach is the combination of the emerging liquid–liquid microextraction procedures (LPME) with TXRF analysis. The great advantage of this procedure compared to conventional liquid–liquid extraction procedures is that equilibrium is reached quickly, owing to the large surface area between the extraction solvent (few microliters) and the aqueous sample (few milliliters), so that the extraction process is almost independent of time. It is a simple, inexpensive, fast, and effective pretreatment technique that has been mostly applied to the analysis of organics and, more recently, for the determination of trace metals in water or biological samples using GFAAS (graphite furnace atomic absorption spectrometry). Taking advantage of the microanalytical capability of TXRF spectrometry, these procedures have been successfully applied in combination with TXRF for the analysis of trace amounts of elements in complex liquid samples such as soil extracts (Marguí et al., 2010a).

- **Solid samples**

As shown in Figure 6.7, the TXRF technique is also used for the analysis of solid samples with or without a previous sample treatment procedure (i.e., dissolution, suspension, decomposition).

The direct analysis of solid samples entails the deposition of minute amounts of sample (usually a few grains) on the sample carrier by means of the end of a clean cotton wool bud (Q-tip®) as a sampling device. In this context, TXRF is a valuable tool as a microanalytical technique for the analysis of scarce or high valuable samples, especially in the fields of forensics and archaeometry. In Figure 6.9, an example of sampling of small particles of an ink written letter from an ancient parchment is shown.

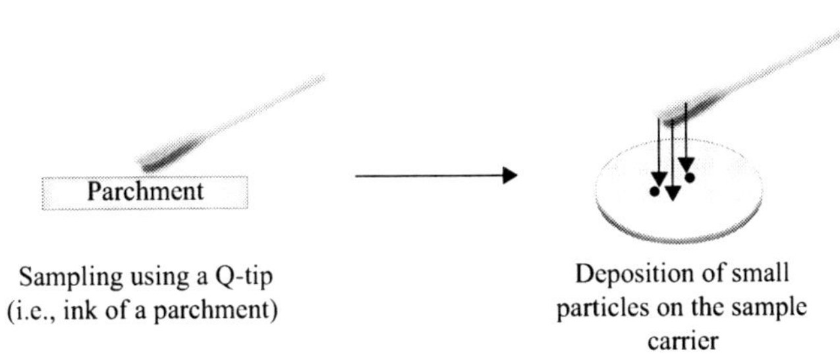

Figure 6.9. Sampling of small particles of an ink-written letter from a parchment for TXRF analysis.

The first step in the delicate sampling process is to rub off carefully, with a cotton swab, some grains from the area of interest. Then the sampled grains that adhered to the cotton wool of the Q-tip must be gently knocked off the Q-tip so that it is deposited onto the sample carrier for subsequent TXRF analysis. For extra fixation, the reflector can be coated with a thin layer of vacuum grease in advance. Considering that in the TXRF configuration, the detector is placed very close to the sample, a sample thickness greater than 0.5 mm must be avoided to prevent damage or contamination of the detector. Usually, this kind of preparation is valid only for determining the qualitative element content of target samples. If a semi-quantitative/quantitative analysis is required, then more sophisticated sample treatments are usually employed.

Another possibility for the analysis of solid samples is the dissolution or suspension of a small amount of sample using an appropriate solvent. If the sample material does not already exist in a finely ground form, grinding down to grain sizes of <75 μm must be performed by using suitable devices (mortar, mills, etc.). Then, 20–50 mg of the sample is usually weighed in a sealable container with a volume of about 1–5 mL of the solvent or dispersant solution. In the latter case, ultrapure water or diluted solutions of commercial nonionic detergents (i.e., Triton X-114) are commonly used. It is important to note that the choice of appropriate suspension concentration (sample/dispersant ratio) is crucial. If the condition of a thin layer is not maintained, the analysis is then not performed under total reflection conditions and we can get detrimental errors in the analytical results.

After the dissolution or suspension of the sample, an appropriate amount of internal standard may be added for internal standardization. After thorough homogenization, a few microliters of the mixture are transferred to a sample carrier for the subsequent TXRF analysis.

For solid samples containing a high amount of organic matter (e.g., biological samples), the separation of the matrix by cold plasma ashing (CPA) can lead to a significant improvement in analytical results. This technique allows for the gentle oxidation of organic matter, which also reduces the thickness of samples for the later TXRF analysis (Woelfl, Mages, and Encina, 2003). In some applications, the use of more complex sample treatment procedures, similar to those employed in atomic spectrometry techniques, is required to achieve limits of detection in the milligram per kilogram to microgram per kilogram range. In this context, wet ashing involving digestion with strong acids has been used to decompose a sample and dissolve the analytes in a wide variety of solid matrices prior to TXRF analysis. In sample dissolution for determining the total contents of elements, the majority of wet digestion procedures require the use of several reagents to achieve the destruction of the organic matter (including nitric, sulfuric, and perchloric acids, and hydrogen peroxide) and/or to ensure the dissolution of inorganic compounds (using hydrochloric acid and hydrofluoric acid) present in a sample. The choice of a procedure has to be optimized as appropriate to the type and nature of the matrix. In all cases, high-quality reagents (i.e., Suprapur® or equivalent quality) must be used to avoid any contamination.

Over recent decades, microwave ovens have been employed extensively instead of conventional sources of heating for shortening the time required for sample digestion and to avoid analyte losses and contamination from other samples or from the surroundings. Finally, it is interesting to remark that in many cases, the resulting solutions after the decomposition procedure have to be concentrated to achieve the required sensitivity. The preconcentration treatment procedures mentioned above (liquid sample analysis) can be used for such a purpose.

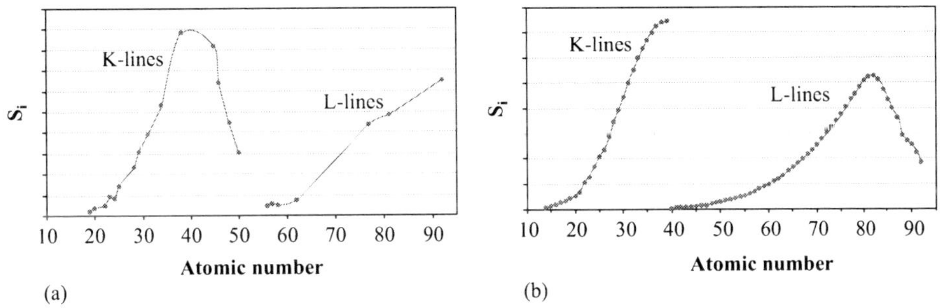

Figure 6.10. Element sensitivities for different TXRF spectrometers: (a) W excitation, (b) Mo-K excitation (S2 PICOFOX, Bruker AXS).

6.3.1.3 Quantification

Taking into account that for TXRF measurements, samples must be prepared as thin layers, then no matrix effects such as absorption and enhancement (see Section 3.3.1) common to WDXRF and EDXRF spectrometry occur. Therefore, it can be assumed that the intensities measured are directly proportional to the mass fractions of the elements of interest and a single set of relative sensitivities can be used for quantitative analysis. It has to be noted that the relative element sensitivities (sensitivities relative to the sensitivity of a particular element) are only valid for a particular system configuration and components, thus, changes of the excitation source or the detector lead to new element sensitivities. Usually, commercial TXRF systems are supplied with a relative sensitivities library linked to the instrument software. In Figure 6.10, examples of relative sensitivities depending on element atomic number for different TXRF configurations (Mo-K excitation and W excitation) are displayed.

As is shown in Figure 6.10, element sensitivities vary over a broad range of values not only because of the fluorescence yields but also instrument-related properties such as the excitation source (e.g., W and Mo excitation) and efficiency of the detector. The sensitivity calibration can be extended for nonmeasured elements which have characteristic energies within the limits given by the characteristic energies of the measured elements.

Based on the element sensitivities, three basic quantification models (described here) can be used.

- **Quantification with an internal standard**

Internal standardization is one of the most commonly used strategies for quantification purposes in TXRF analysis. It is based on the addition of a defined amount of a liquid monoelement standard (internal standard) to the target sample. Then the concentration of the respective elements in the sample can be determined from the following equation:

$$C_i = \frac{C_{IS} N_i S_{IS}}{N_{IS} S_i} \tag{6.2}$$

where C_i is the concentration of the element to be analyzed, C_{IS} is the concentration of the internal standard, N_i is the net signal of the analyte, S_{IS} is the relative sensitivity of the internal standard element, N_{IS} is the net signal of the internal standard element, and S_i is the relative sensitivity of the analyte.

To select the type and concentration of the internal standard, several characteristics have to be taken into account:

- The internal standard is not present in the sample (usually Ga, Y, or some rare earth element is used).
- The fluorescence lines of the internal standard should not interfere with those of the elements of interest in the sample.
- The concentration of the internal standard should be within the average concentration range of the elements of interest.

This quantification approach can be applied to liquid samples, digested samples, and suspensions. In addition to quantitative analysis purposes, the use of an internal standard reduces inaccuracies introduced during sample deposition onto the sample carrier.

- **Standardless quantification**

For microanalysis of powders, single grains, or fibers that must be deposited onto the sample carriers, weighing of the minute amount in micrograms is not practically feasible. In such cases, the concentration of elements with respect to this small mass cannot be determined. Then, the analyte concentration can be calculated as

$$C_i = \frac{N_i}{S_i \sum_{1}^{n} \frac{N_i}{S_i}} (100\% - C_B) \tag{6.3}$$

where C_i represents the concentration of the element to be analyzed, N_i is the net signal of the analyte, C_B represents the concentration of the sample portion not detected and S_i is the relative sensitivity of the element to be analyzed.

This method of quantification allows for the quantification without the addition of an internal standard but its application is limited to relatively few known sample types (e.g., alloys). For example, a powdered brass alloy (containing Cu and Zn) with impurities of Fe and Pb could be analyzed by this method.

- **External calibration**

Element quantification by TXRF can also be performed by means of external calibration using a set of standards. In this case, the absolute element signal is plotted versus known concentration of the element of interest in the standards and a straight line is fitted to measured points by the least-squares linear regression method. Then, the element concentration in the target sample is calculated by interpolation from the developed calibration curve.

This approach is usually selected when a preconcentration treatment is employed before the TXRF analysis. Usually, the recovery of the preconcentration procedure is not 100% and, as a result, direct internal standardization for quantification purposes cannot be used.

6.3.2 SURFACE ANALYSIS

In addition to the chemical analysis capabilities of TXRF systems (discussed in Section 6.3.1), TXRF can also be applied to surface and near-surface layer analysis.

In the region of total reflection, the fluorescence intensity is highly dependent on the glancing angle of incidence. By measuring the angle dependence of the fluorescence signal, the location of the fluorescing atoms above or below the surface becomes possible. In Figure 6.11, an example of the curves measured for a residue on the surface, a surface layer, and a bulk sample are compared. These show that by running an angle scan, it is possible to discriminate between trace impurities that are homogeneously distributed in thick samples from those located in granular particles on the carrier or evenly deposited in thin near-surface layers.

This is the main reason why TXRF is widely used within the semiconductor industry as a routine contamination control system for the analysis of Si wafer surfaces. In such an application, the sample is the wafer itself which has a polished plane surface, as required for total reflection of X-rays.

As already stated in the previous section, to carry out this kind of measurement a special TXRF setup allowing measurements at various angles is required. The angle-dependent intensity profiles are used for qualitative and quantitative characterization of the samples under investigation. Moreover, the possibility to map (scan several points on the surface and determine the level of contamination as a function of the spatial distribution) of a wafer can help locate and characterize sources of contamination. For quantitative analysis, in this case, the use of internal standardization is not recommended to avoid any possible further contamination. Therefore, the calibration is performed using external standards. Most of the TXRF systems in use for this application are calibrated at a single point with a Ni droplet residue (1 ng) on a reference wafer.

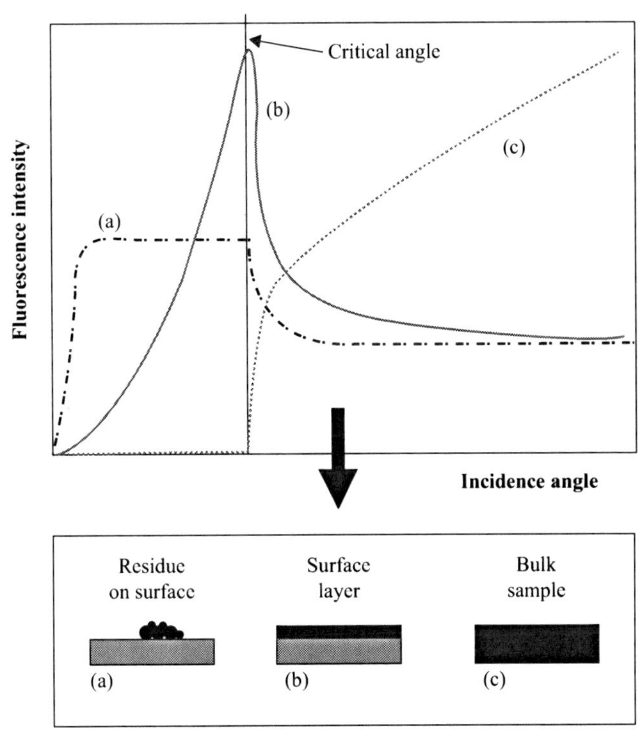

Figure 6.11. Angle dependence of the fluorescence signal for various locations of an analyte (Modified from Streli (2006)).

The detection limit of the direct nondestructive TXRF measurement of Si wafers is about 10^9 atoms/cm^2 (for the critical transition metals). However, lower limits of detection (10^7 atoms/cm^2) can be achieved if the impurities over the entire surface of the wafer are collected and preconcentrated by vapor-phase decomposition (VPD) prior to the TXRF analysis. In VPD, the surface of the wafer is exposed to hydrofluoric acid vapor that causes the surface oxide to dissolve together with the impurity metals. The acid droplets, condensed on the surface, are then collected to the center, dried, and analyzed by TXRF at a fixed-angle setting.

Another interesting surface analysis application of TXRF is for the characterization of thin-layered materials and implantations by using angle dependent XRF analysis in the region around the critical angle of total reflection. When the angle of incidence passes through the critical angle, the penetration depth of the incoming X-rays changes from the nanometer to the micrometer region and thus, information can be obtained on depth distributions and thin films over that range. If the analyzed layer materials are perfectly homogeneous and flat, the number and sequence of the individual layers, their element composition, thickness, and density can be determined from a qualitative interpretation of the TXRF element profiles. However, a deep understanding of the layered structures and recorded profiles is required and thus this kind of analysis is limited to skilled operators with considerable experience.

6.4 OTHER APPLICATIONS OF TXRF AND CASE STUDIES

As stated in the previous section, taking advantage of its inherent characteristics, TXRF has been widely applied for both chemical and surface analysis in many fields of study. In this section, the main applications of both types of analyses are summarized. Moreover, at the end of this section, some case studies illustrating the possibilities and drawbacks of TXRF are presented.

As shown in Table 6.3, TXRF has been applied for chemical analysis of many different kinds of samples (Van Grieken and Markowicz, 2002). Usually, the simultaneous multielement capability, the low detection limits for many elements, and the easy calibration at a relatively low cost have promoted its use in many studies.

In the environmental field, TXRF has been mostly applied for trace element determination in different types of water samples. Because the background is so low, element concentrations in the range of a few tenths of a microgram per liter can be obtained in aqueous solutions such as drinking water and river water. However, when dealing with the analysis of waste and sea water, additional sample preparation methods have to be applied to separate the suspended matter and to remove the salt content prior to the TXRF measurement (see Section 6.3.1.3 for details).

Aerosol analysis is also an ideal application for TXRF as only sample amounts of several nanograms are required for analysis. An additional advantage of TXRF is that aerosol samples can be sampled directly onto the sample carrier and thus, sample preparation is reduced to a minimum.

Environmental monitoring can also be undertaken by analyzing soil, sediments, and different vegetation and animal species. In such cases, a proper sample preparation of the sample (i.e., sample decomposition) is needed before the TXRF analysis.

TXRF has been widely applied for the analysis of medical and biological samples because trace elements have an important biological function and impact on all living beings. Generally, body fluids are recommended samples for TXRF analysis and numerous studies based on the

Table 6.3. Examples of some of the main applications of TXRF for chemical analysis

Field	Type of sample
Environment/Geology	Water (rain, river, sea, drinking, waste) Air (aerosols, airborne particles, dust, fly ash) Soil, sediment, sewage sludge Ore rocks, minerals Vegetation (leaves, needles, roots, wood) Fauna (fish, mussels)
Medicine/Biology/Pharmacology	Body fluids (blood, serum, urine, amniotic fluid) Tissue (hair, kidney, liver, lung, nails) Various (enzymes, glucose, proteins, biofilms)
Industrial/Technical	Chemicals (acids, bases, salts, solvents) Oil (crude oil, fuel oil, grease) Beverages (wine, tea, alcoholic drinks)
Fine arts/Archaeometry	Pigments, paintings, varnish, pottery, bronzes
Forensics	Textile fibers, gunshot residue, drugs, fingerprints

analysis of body fluids and biological tissues have been published during the last few decades. When dealing with this type of sample, usually a decomposition procedure or the formation of a solid suspension is preferred as the sample preparation procedure.

Quality control of ultrapure reagents, like acids, bases, and solvents can also be performed by TXRF. Another interesting application in the industrial field is the analysis of oil samples. These are difficult to analyze with other atomic spectroscopic techniques owing to their inherent viscosity. For TXRF analysis, oil samples can be directly deposited onto the sample carrier with or without a previous dilution using an organic solvent (e.g., toluene or chloroform). Using such an analytical approach, limits of detection in the range of micrograms per milliliter are achievable. If detection limits of nanograms per milliliter are required, an additional sample treatment based on low temperature oxygen plasma ashing is necessary. The analysis of liquid foodstuffs (e.g., wine, alcoholic drinks) by TXRF is another application field of industrial importance.

As already stated in Section 6.3, TXRF has proven to be a useful technique for the analysis of small amounts of solid samples. In this context, TXRF has become a valuable microanalysis tool to study rare or high valuable samples in the field of fine arts and archaeometry. For instance, many analyses of microsamples have been made for resolving questions concerning objects of cultural heritage (e.g., their provenance, detecting fakes, comparison of artifacts, detection of unknown materials, etc.). In this category are included the determination of the chemical composition of a great variety of pigments and paintings, inks used for the production of ancient manuscripts, musical instrument varnishes, and glass objects.

For forensic applications, the microanalytical capability of TXRF has particular importance as only microsamples are usually available. In recent years, several specific problems have been investigated, including the analysis of hair samples, drug powder, gunshot residues, and textile fibers.

For surface analysis, as discussed earlier, TXRF has become a successfully established quality control method in the semiconductor industry, particularly for the ultratrace element analysis of Si wafer surfaces. TXRF applications can fulfill general industrial requirements of daily routine monitoring of wafer cleanliness up to 300 mm diameter under clean room conditions. Elements such as S, Cl, K, Ca, Ti, V, Cr, Mn, Fe, Co, Ni, Cu, Zn, As, Br, Sn, Sb, Ba, and Pb can be directly monitored at trace levels (~10^9 atoms/cm^2). For the determination of low-Z elements (i.e., Na, Al, and Mg), TXRF has also been extended but its implementation for routine analysis needs further research. Novel detector systems, excitation sources, and automated sample preparation techniques are being developed to cope with this extended need (Pahlke, 2003).

A variety of thin layers deposited on flat surfaces have been also characterized by TXRF, for example, thin-layered materials with layer thickness in the range of nanometers used in, for example,integrated cricuit technology and the glass industry.

So far, we have presented the most common applications of TXRF systems using X-ray tubes as excitation sources. However, in recent years, the combination of TXRF with synchrotron radiaton (SR) (see Chapter 7 for details) has extended the applications of TXRF. With the use of SR-TXRF setups, the limits of detection for element determination are significantly improved (down to the femtogram level). The combination of TXRF with synchrotron radiation also allows for a combination with atomic absorption spectroscopy, which extends chemical speciation to the trace element level. However, the access to large-scale storage rings for electrons and positrons, necessary SR-TXRF systems, is restricted and these beam lines are usually employed only for more fundamental research purposes. An overview of recent advances in this field can be found in a recent review (Streli, 2006).

To illustrate some typical applications of TXRF used today, the determination of trace amounts of metals in waste water effluents, in geological samples, and on Si wafer surfaces have been selected and are described below. Some guidelines for laboratory work, including sample preparation and quantification, are also discussed in each of these sections.

6.4.1 MULTIELEMENT DETERMINATION IN WASTE WATER EFFLUENTS

The monitoring of heavy metals in industrial waste water effluents is an important activity in many laboratories. Of special interest is the screening of the elemental composition (As, Ba, Cd, Cu, Cr, Sn, Fe, Mn, Ni, Pb, Se, and Zn) of inlet effluents and quantitative analysis of outlet effluents to study the efficiency of chemical treatment processes to eliminate metals and to comply with current statutory legislation concentration limits. In this context, fast analytical methodologies that entail simple sample preparation are preferable.

Because of the high amounts of suspended solids and the viscosity of some industrial waste water samples, an experimental evaluation of the influence of the sample volume used on TXRF results has to be performed to ensure that the sample is presented as a thin film. For routine and screening analysis of industrial inlet and outlet effluents, TXRF waste water samples analysis can be directly performed by depositing 20 μL of the internal standardized sample onto a quartz reflector and using a measuring time of 1000 s (Marguí et al., 2010c). Using such conditions, the accuracy of the concentration values obtained by internal standardization (using Y as internal standard) is, in general, acceptable, with values determined typically within ca. 10% of the true concentration value. Relative standard deviations for the analysis of three independent replicates also proved satisfactory with values for all elements <10%, and mostly <5%. A further enhancement of the analytical quality of TXRF results can be achieved using more sophisticated sample treatments such as microwave acid digestion.

The fact that the TXRF analysis may be performed directly on the raw waste water entails less sample manipulation and requires lower amounts of reagents and therefore, cost and time savings are achieved for the industrial laboratories. Additional advantages of the TXRF method are the multielemental information obtained for the sample, the easy quantitation through internal standardization, and also low operating costs if a benchtop system is used (they do not require cooling media and gas consumption for function).

All these benefits make this analytical technique very attractive for implementation in the industrial treatment plants industry to increase the productivity of the laboratory; it offers an important and complementary analytical tool to the digestion procedures coupled with atomic spectrometry for the analysis of these types of waste water samples.

6.4.2 DETERMINATION OF TRACE AMOUNTS OF Se IN SOIL SAMPLES

Selenium concentrations in soils are an important issue owing to the narrow range between the nutritious requirement and toxic effects upon Se exposure for humans. However, its determination is challenging owing to the low concentrations present within complex matrices that hamper the analysis for most spectroscopic techniques.

The use of direct analysis of soil suspensions by TXRF could be interesting as a fast and relatively simple methodology to get a first idea of the multielemental composition of the soil. Moreover, it could be successfully applied to determine the Se content when only an extremely small amount of sample is available (only 20 mg of soil required) provided that the Se content is sufficiently high. A further improvement of the Se detection limit can be achieved using a more sophisticated sample treatment on the digested soil sample. However, it has to be kept in mind that when using additional sample treatments focused on Se determination, the inherent TXRF multielemental information on soil composition becomes no longer possible. The Se detection limit for the direct analysis of the soil digest by TXRF was found to be 18 μg/L corresponding to 0.76 mg/kg in the soil (Marguí et al., 2010a). Despite the improvement on Se determination compared to the analysis of soil suspensions, the detection limit achieved is not adequate for most environmental applications (the worldwide average concentration of Se in soils is ~0.4 mg/kg). Better results in terms of sensitivity can be achieved when decreasing the Fe concentration in the soil digests (~700 mg/L) by the application of a liquid–liquid extraction procedure using ethyl ether or simply by the introduction of a Cr absorber into the instrument configuration. Using both approaches, the detection limit for Se is reduced by around 35% compared to the direct analysis of the soil digest. However, to get accurate results, Se concentrations in the milligram per kilogram range have to be present in the soil. Therefore, environmental applications are similar as for the solid suspensions.

On the other hand, if the Se content in the soil samples at or below the milligram per kilogram level has to be determined, an additional sample preparation step is required. The use of a liquid phase microextraction (LPME) procedure to extract the Se content from the soil matrix together with the microanalysis capability of TXRF spectrometry to thereby improve the detection limit for Se is a good alternative. The main advantages of the proposed method include the low consumption of solvents, low cost, speed, and simplicity of operation. With the LPME sample preparation, a Se detection limit of 1.1 µg/L in the soil digest corresponding to 0.05 mg/kg in the soil was achieved with only 2 mL of soil digestion solutions, making it suitable for the application to environmental samples. Using this simple and low cost sample preparation strategy, accurate results can be obtained at the low milligram per kilogram range and the Se detection limit is almost 10 times lower than the worldwide mean concentration of Se in soils, indicating that the proposed method combining LPME and TXRF analysis can be suitable for such determinations. Moreover, the Se detection limits achieved when the LPME procedure was applied before the TXRF analysis are in most cases much better than those reported in other published works for the determination of Se in soils using XRF spectrometry or INAA (Instrumental Neutron Activation analysis) and are also competitive with those determined using other popular spectrometric techniques such as GFAAS, ICP-MS, and AFS (Marguí et al., 2010a).

6.4.3 ANALYSIS OF Si WAFER SURFACES

Ultraclean Si wafers are very critical to ultralarge scale integration (ULSI) manufacturing because uncontrolled contamination during the wafer processing can change their electrical characteristics, resulting in yield loss. Sources of metal contamination include processing chemicals used during cleaning, atmospheric deposition, processing equipment, and wafer handling. Currently, TXRF analysis is widely used in the semiconductor industry as a routine contamination control system for the analysis of Si wafer surfaces. As was stated in Section 6.3.2, the calibration is performed using external standards, and as mentioned already, for quantification purposes, most of the TXRF systems are calibrated at a single point with a Ni droplet residue (1 ng) deposited onto a reference wafer. Ni is usually selected because this element is not a contaminant of Si wafer surfaces. However, the shape of the droplet residue is a critical issue that highly influences the quality of the final results. The comparison of different theoretical samples shapes (cuboid, cylinder, spherical segment, and ring shape) has shown that the ring shape is the best option to obtain the highest fluorescence intensity and the lowest saturation effect. It has been demonstrated that a large diameter, a small width, and a small height lead to the best results (Horntrich et al., 2012).

Direct surface measurements by TXRF usually suffer from a lack of sensitivity, and additional sample treatments (i.e., VPD, see Section 6.3.2 for details) are used to enhance detection limits. However, some elements (such as Cu) may be problematic at certain concentrations and thus an evaluation of recovery for the target elements of interest must be performed and established to report good quality, meaningful analytical results (Wang et al., 2000).

CHAPTER 7

SPECIAL XRF CONFIGURATIONS AND RELATED TECHNIQUES

7.1 INTRODUCTION

In addition to the more commonly used X-ray tubes and to a lesser extent the radioisotopes, discussed in detail in Section 2.2 in Chapter 2, there are other sources used in X-ray analysis based on the use of accelerated particles. Of these the most important ones are those that use synchrotron radiation, particle-induced X-ray emission, and electron-induced X-ray emission. Except for the latter case, these techniques are rarely available to the general user community, as large electron or other charged particle beams source facilities are needed—although, they cover interesting, important applications in many fields. In addition, considerable advances and improvements in X-ray optics have also resulted in improved analytical performance leading to new applications of XRF at the micrometer scale.

In the following sections, the basic principles, characteristics, and main applications of XRF variants based on the use of accelerated particles and/or the use of focusing optics devices is presented.

7.2 MICROBEAM X-RAY FLUORESCENCE SPECTROMETRY (μ-XRF)

Microbeam X-ray fluorescence (μ-XRF) is one of the most recent branches of XRF spectrometry, which has developed rapidly since the development of capillary optics in the early 1990s. The main particularity of μ-XRF instrumentation compared to conventional XRF systems is the focusing of the X-ray beam down to a much smaller size by introducing appropriate focusing optics, usually monocapillary or polycapillary lens, between the X-ray source and the sample (for additional information see Section 2.5.3 dedicated to focusing optics). Standard microbeam spectrometers are also usually equipped with a conventional microscope (or just a camera) for a precise selection of the measurement spot and for viewing the area on a PC monitor during the alignment and measurement procedures. Moreover, the use of an *xyz*-programmable motorized stage enables fully automatic measurements to be made and the possibility for performing the fluorescence spectroscopy in different modes: point analysis, line scan, or using an area scan. In the period 2002–2004, a new variant of μ-XRF system based on the use of two focusing devices was developed. In such systems, called *confocal XRF spectrometers*, a first focusing device is introduced between the source and the sample (like in conventional μ-XRF) and a second focusing device is introduced between the sample and the detector to restrict the volume in the sample from which the XRF radiation can reach the detector. Therefore, by moving this sampling volume within a material, it is also possible to record element X-ray intensity profiles versus depth.

The main analytical capabilities associated with μ-XRF are:

- **Obtaining information on sample composition and related spatial element distribution at the micrometer (or even nanometer) scale.** This kind of information is particularly valuable for studying element distributions within heterogeneous samples. When capillary optics are used, the commercially available μ-XRF instruments using X-ray tubes as radiation sources allow for resolutions between 10 and 100 μm and acceptable sensitivities (at nanogram level). However, a significant improvement in terms of spatial resolution and sensitivity can be achieved if more powerful sources based on the use of accelerated particles are employed. Within these the most important one is the use of synchrotron radiation (for more detail see Section 7.3), which allows one to get elemental information at the nanometer scale with sensitivities at the picogram or femtogram level.
- **Possibility to record 2D or 3D elemental maps of the sample surface.** As already stated in the introduction to this section, in μ-XRF systems the sample is placed on an *xyz*-programmable motorized stage that enables fully automatic measurements to be made, such that there is the possibility to perform analysis of a specific area on a sample. Therefore, two-dimensional (μ-XRF systems) or three-dimensional (via confocal XRF systems) elemental maps can be recorded. As an example, in Figure 7.1, 2D-spatial distribution false-color maps of As in rat kidney slices determined by using an X-ray tube-based μ-XRF instrument and by synchrotron μ-XRF (μ-SRXRF) are shown. Each image indicates the relative distribution of the specific element, thus, the concentration scales vary for each image. The normalized X-ray fluorescence intensities are scaled to red, maximum, and to blue, minimum. As can be seen, information on As accumulation patterns in rat kidney can be visualized. In this study, it was found that As tends to accumulate in the renal cortex (nephrons). From Figure 7.1, it can also be deduced that when using a μ-SRXRF system, it is possible to obtain information with a higher spatial resolution in a lower measurement time.

As discussed in Chapter 3, in principle, the XRF technique can be used for quantitative analysis with different calibration approaches. However, quantitative analysis when using μ-XRF is more complex compared to that when using conventional XRF instrumentation because:

- the heterogeneous nature of calibration standards and unknown samples are at the micrometer/nanometer scale;
- the analytical signal obtained is a depth-averaged signal as the primary beam penetrates relatively deeply below the surface of a sample.

As a consequence, for most applications, quantitative results obtained by μ-XRF are associated with relatively large errors and are best considered only as semiquantitative. Therefore,

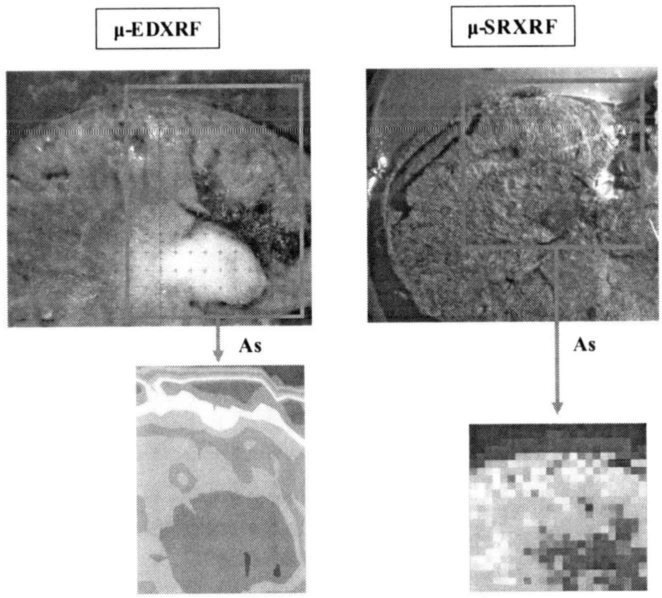

Analytical conditions:

- **μ-EDXRF:** W X-ray tube, 50 kV, grid of 20 × 20 points, monocapillary of 1000 μm diameter, Si(PIN) detector of 190 eV at 5.9 keV, measurement time: 300 s/step.
- **μ-SRXRF:** LNLS Campinas, Brazil, electron bunch 1.37 GeV, critical energy 2.08 keV, orthogonal slits 300 μm × 300 μm, Si(Li) detector of 165 eV at 5.9 keV, measurement time: 100 s/step.

Figure 7.1. Top: photos of cross-sections of rat kidney. Bottom: 2D-maps of As distribution in rat kidney slices by μ-XRF and μ-SRXRF. False-color maps/images are presented as XRF intensities scaled maximum to minimum (modified from Pérez et al. (2006)).

significant effort is being devoted to develop new irradiation/detection and associated quantification strategies or to produce and characterize new calibration standards suitable for μ-XRF so that more reliable quantitative results may be obtained (Janssens et al., 2010). For instance, some researchers have proposed the use of microdots of standard solutions on a suitable substrate for calibration purposes of microanalytical techniques such as μ-XRF. In the application of confocal μ-XRF setups, multilayered polymer standards, in which each layer has a known concentration of specific metals, have been prepared to validate the quantification model for the determination of stratified materials by 3D μ-XRF.

As stated earlier, μ-XRF is usually employed to obtain information on the local elemental composition of homogeneous and nonhomogeneous samples in different areas of study. The choice of using a laboratory μ-XRF system or μ-SRXRF depends on the spatial resolution and sensitivity required for the target analysis. For instance, nanometer scale resolution using μ-SRXRF systems is clearly needed to better understand accumulation pathways of toxic elements in nature-related sample (e.g., interfaces between soil and roots). By contrast, to study metal distribution in contaminated soils, micrometer scale resolution is often sufficient and laboratory-based systems are usually adequate.

Whatever the need, both laboratory μ-XRF or μ-SRXRF systems are being actively employed in many areas of science and technology for a variety of purposes. Some examples are presented in Table 7.1.

Table 7.1. Main fields of application of μ-XRF systems

Field	**Applications**
Environmental and earth science materials	- Particulate analysis - Vegetation tissues analysis (tree rings, roots, etc.) - Metal distribution in soil/sediment profiles
Cultural heritage materials (art and archaeometry)	- Trace element mapping in corroded ancient glasses - Local analysis of bronze statues, decoration on porcelain vases, document forgeries
Biological, medical, and forensic applications	- Analysis of hair, bones, biological tissues - Gunshot residue analysis
Industrial materials	- Plating thickness (control film thickness) - Microelectronics industry (thin film technology) - Waste characterization

7.3 SYNCHROTRON RADIATION-INDUCED X-RAY EMISSION (SRXRF OR SRIXE)

Synchrotron radiation (SR) is the name given to the radiation given off when charged particles are accelerated in a curved path or orbit (circular accelerator or synchrotron).

To produce SR, a source of electrons or positrons are accelerated in a linear accelerator to produce energies around 100 MeV; the resulting beam is then injected into a circular accelerator (synchrotron) and boosted to energies in the giga electronvolt range. Finally, the beam is stored in a separated storage ring to be used at the different cutting-edge experimental stations, which usually consist of an optics hutch, an experimental hutch, and a control cabin.

It is important to realize that to produce energies in the giga electronvolt range, high-power accelerators that have diameters of hundreds of meters are required, thereby costing tens or hundreds of millions of dollars to construct. For this reason, at present, a limited number of synchrotrons facilities are available around the world and these installations are mostly funded by governments and operated as national infrastructure facilities that are available to scientists from universities and research organizations from different countries. Table 7.2 is a list of synchrotron facilities sited in different countries (with their associated radiation energy).

In an SR facility, the beam is stored in a separate storage ring to be used in different experimental stations named *beamlines*. The selection of a beamline by the users is usually made taking into account the beam description, the energy range, and the beam size. As an example, in Figure 7.2, a graphic view of the available beamlines at the European Synchrotron Radiation Facility (ESRF) is shown. A more detailed description of each beamline can also be found at the website of the SR facility (see Table 7.2).

As a limited number of synchrotron facilities are available globally the use of these installations is strictly controlled and a formal application for beamtime must be submitted. Standard

Table 7.2. List of some synchrotron facilities around the world

Synchrotron source	Location, Country	Energy (GeV)	URL
		Europe	
ALBA	Barcelona, Spain	3	www.cells.es
ANKA	Karlsruhe, Germany	2.5	www.ankaweb.fzk.de
ASTRID	Aarhus, Denmark	0.58	www.isa.au.dk
BESSY II	Berlin, Germany	1.5–2.0	www.bessy.de
DIAMOND	Oxfordshire, England	3	www.diamond.ac.uk
ELETTRA	Trieste, Italy	1.5–2.0	www.elettra.trieste.it
ESRF	Grenoble, France	6.0	www.esrf.fr
MAXII	Lund, Sweden	1.5	www.maxlab.ln.se
SLS	Villigen PSI, Switzerland	2.4	www.psi.ch/sls
SOLEIL	Paris, France	3	www.synchrotron-soleil.fr
		Asia	
BSRF	Beijing, China	1.4–1.55	www.ihep.ac.cn
INDUSII	Indore, India	2.5	www.cat.ernet.in
PLS	Pohang, Korea	2.0	www.paleng.postech.ac.kr
Spring-8	Hyogo Prefecture, Japan	8.0	www.spring8.or.jp
SSRF	Shanghai, China	3.5	www.ssrf.sinap.ac.cn
		America	
ALS	Berkeley, USA	1.0–1.9	www-als.lbl.gov
APS	Argonne, USA	7.0	www.aps.anl.gov
CLS	Saskatoon, Canada	2.5–2.9	www.lightsource.ca
LNLS	Campinas, Brazil	2.0	www.lnls.br
		Australia	
Australian synchrotron	Melbourne, Australia	3	www.synchrotron.org.au

templates for beamtime requests are available on SR facilities' websites. Usually, a proposal should include:

- proposal title (including the research area of the proposal);
- beamline requested (beamtime analytical characteristics: energy range, beam size);
- Number of shifts required to carry out the project (requests for beamtime are based on shifts of 8 h);

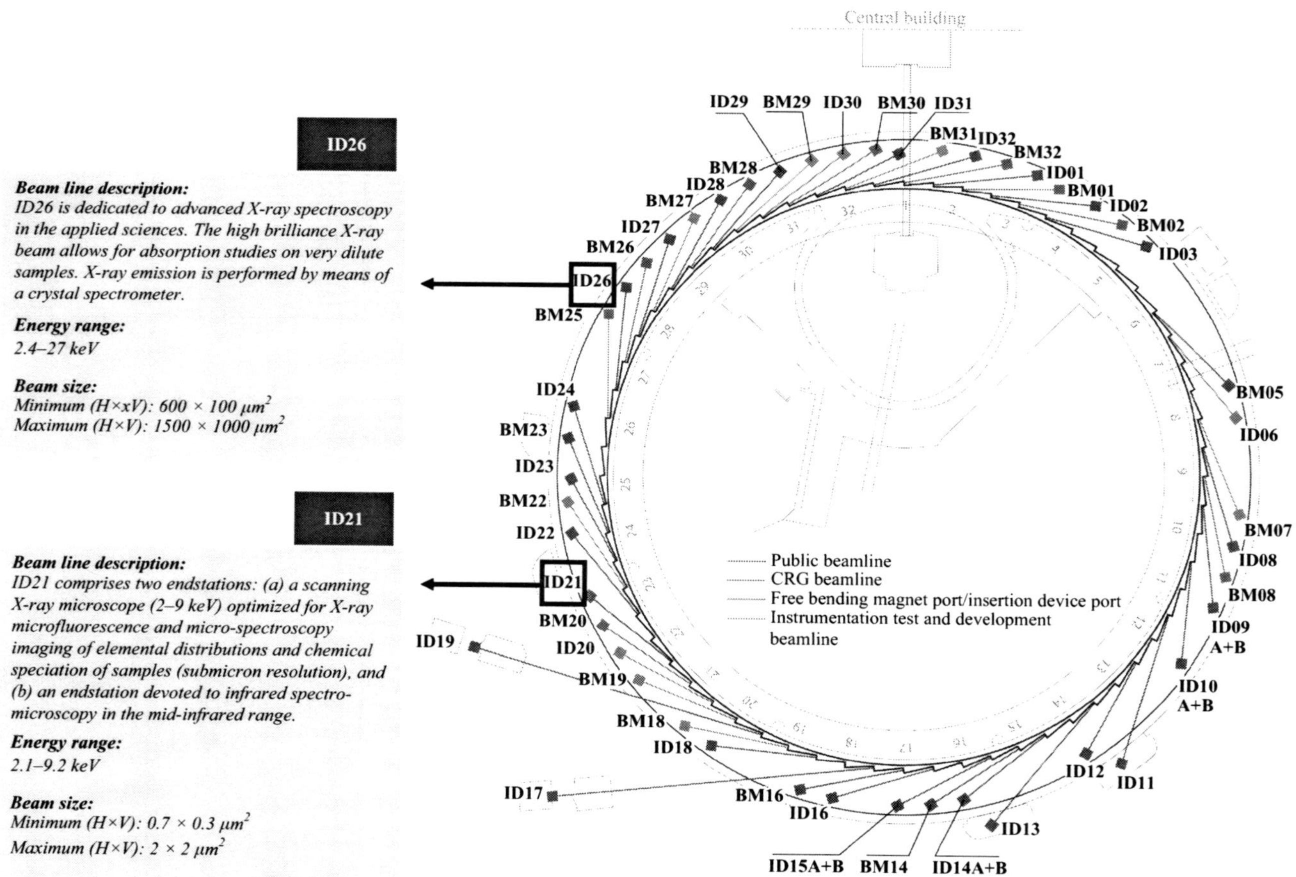

Figure 7.2. Beamlines available at the ESRF (Grenoble, France) (Modified information from the ESRF website (www.esrf.fr)).

- main proposer (to whom correspondence will be addressed) and coproposers;
- laboratory support facility (if an additional chemistry or biology laboratory is needed to prepare the samples before the SRXRF analysis);
- sample environments (if additional devices are needed during the measurement process: furnace, refrigerator, cryostat, etc.);
- sample description (mass, size) and safety (if the sample is radioactive, biologic, explosive, etc.);
- technical reasons which make the beamline requested necessary for the target experiment;
- experience with synchrotron radiation (including scientific research papers published using SRXRF).

Usually, two proposal review rounds are held annually. The assessment of beamtime is allocated following the recommendations of a review committee that is composed of specialists in relevant areas of science.

From an analytical point of view, the synchrotron produced X-ray beams have unique properties that make them desirable for particular uses. These properties are:

- **High intensity, monochromatic beam which is highly polarized.** The intensity of synchrotron radiation is four to five orders of magnitude greater than that of conventional X-ray tubes. This, in combination with it being monochromatic and with a high degree of polarization in the plane of the synchrotron ring, allows for assessing elemental information on a sample with absolute detection limits at the femtogram level (compared to the microgram or nanogram level usually associated with conventional laboratory-based XRF systems).

  The geometric arrangement of SRXRF beamlines is similar to conventional EDXRF systems but in which the primary source is not an X-ray tube but SR. However, because of the analytical capabilities considered already, SR has also been successfully applied in combination with other XRF configurations including: μ-XRF (for details see Section 7.2), TXRF (for details see Chapter 6), XAS (X-ray absorption spectrometry), and XRD (X-ray diffraction).
- **μ-SRXRF (Microsynchrotron radiation-induced XRF).** As the SR beam is high intensity, very rapid bulk analysis at trace and ultratrace levels can be undertaken even on small areas of a sample. For this reason, this sort of radiation source has been widely applied in combination with μ-XRF, as already discussed in the previous section (Section 7.2). For instance, absolute detection limits of <0.1 fg have been obtained using a 21 keV monochromatic radiation at 2 μm × 2 μm in a live time measurement time of 100 s (Somogyi et al., 2001). SR-based μ-XRF offers a number of advantages compared to other microprobe techniques: it combines high spatial resolution with high sensitivity, it can be used in atmospheric conditions, and it causes relatively little beam damage to the sample (this fact is important when analyzing biological and organic materials that are more susceptible to beam-induced damage). Moreover, two-dimensional elemental maps with appropriate sensitivity and spatial resolution can be recorded when using μ-SRXRF systems as shown, for example, in Figure 7.1.

- **SR-TXRF (Synchrotron radiation-induced total reflection XRF).** SR has also been used in combination with TXRF. Very good detection limits have been reported employing this arrangement for the analysis of small quantities of aqueous solutions dried onto a reflector with detection limit in the femtogram range for most elements. SR as an excitation source offers the advantage of an intense highly collimated X-ray beam covering a wide spectral range, including the low-energy region. Therefore, it is the ideal source for the determination of light elements by TXRF. This is particularly important in the quality control of Si-wafer surfaces. In the semiconductor industry, Al and Na are of special interest (they influence the production yield of integrated circuits negatively) and it is of importance to have analytical techniques with adequate sensitivity to detect these elements at ultratrace levels. In addition to the analytical aspects, SRTXRF is also being used in combination with angle-dependent measurements or X-ray absorption spectrometry (see following section) to gain additional information about investigated samples. Supplementary information about the fundamentals of SRXRF and recent industrial and environmental applications can be found in a recently published review (Meirer et al., 2010).
- **SR-XAS (Synchrotron radiation-induced X-ray absorption spectroscopy).** SR can be also combined readily with X-ray absorption spectroscopy (XAS). The fundamentals of XAS are based on irradiation with a monochromatic X-ray beam of tunable energy (here SR) and scanning over an absorption edge of an element of interest while recording the absorption of the beam or the fluorescence radiation produced. From an analytical point of view, by using XAS, information on the chemical environment of the element (speciation) can be assessed. This information is of significance in many fields as the toxicity and chemical behavior of most elements is highly dependent on their oxidation state.

  Recently, the combination of μ-SRXRF and SRTXRF with XAS has become an important analytical tool for speciation in environmental, geological, and industrial materials and for the study of processes in chemical species transformations. In general, μ-SRXRF is used to locate the area of interest and μ-XAS to gain insight into the chemical bonding and the oxidation state of the elements of interest (Fittschen and Falkenberg, 2011).
- **SR-XRD (Synchrotron radiation-induced X-ray diffraction).** SR has also been used in combination with X-ray diffraction (XRD). XRD is based on constructive interference of monochromatic X-rays with a crystalline sample (according to Bragg's law). From an analytical point of view, XRD is primarily used for phase identification of a crystalline material or to provide information on unit cell dimensions. The main applications of XRD in combination with μ-SRXRF are in obtaining μ-XRD maps (with the repartition of crystallographic states) together with elemental information maps. This combined information can be of great significance for the characterization of complex samples.

To sum up, at present, SR as a source for XRF and associate techniques is an essential tool to enhance research activities for a large number of disciplines. Because of their sensitivity and spatial resolution, the synchrotron-based techniques enable one to determine trace and ultratrace amounts of elements within complicated matrices at the molecular and nanometer scale. Several surveys of applications of SR have recently been published (Meirer et al., 2010; Fittschen and Falkenberg, 2011). A few examples of different applications are given in Table 7.3.

Table 7.3. Main fields of application of SRXRF systems

Field	Applications
Archaeology	- Low resolution measurements on different materials (coins, paintings, ancient objects, etc.)
Biology/Medicine	- Determination and spatial distribution of essential and toxic elements in different biological tissues: bones, brain, and so forth - Microanalysis of single cells
Geology/Environmental	- Small particle analysis (fly ashes, aerosol particles) - Determination and spatial distribution of elements in vegetation tissues to study uptake of metals in contaminated soils and waters. - Analysis of thin sections of rocks for studies of specific types of minerals in heterogeneous matrices
Materials Sciences	- Localized corrosion of stainless steel - Measurement of supported catalysts - Determination of source contamination in Si-wafer surfaces through analysis of the chemical elemental state of impurities

7.4 PARTICLE-INDUCED X-RAY EMISSION (PIXE)

Despite the fact that the particle-induced X-ray emission (PIXE) technique is considered a variant of XRF, such a classification is not strictly correct as it is not based on excitation of the sample by electromagnetic radiation. In PIXE, bombardment with ions of sufficient energy (usually megaelectron volt protons) causes inner shell ionization of the atoms in the sample and the subsequent decay of the induced electronic states results in emission of X-rays (that are usually measured by an ED detection system). Therefore, as in XRF systems, one can get multielemental information of the target sample. However, since its development, the application of PIXE has been limited mostly to physics and nuclear research laboratories given that a particle accelerator is needed to obtain the incident proton beam of 1–4 MeV and thus, commercial instrumentation is not commonly available.

From an analytical point of view, the PIXE technique has unique properties that make it desirable for use:

- **High sensitivity and low detection limits.** The high sensitivity of the PIXE technique compared to systems using other sources (i.e., X-ray tubes) is largely owing to the low background in the resulting spectrum. Moreover, compared to X-rays, protons have the advantage that they can be focused by electrostatic or electromagnetic lenses and may be transported over large distances without loss in beam intensity. Therefore, the number of impinging particles per square centimeter per second is higher in PIXE than in conventional tube-excited XRF systems. Taking into account these properties, detection limits for most elements are usually better in PIXE, particularly for organic matrices. If thin targets of the sample are analyzed, limits of detection lower than <1 μg/g can be achieved for most elements.

- **Microanalytical capability.** Focusing of the proton beam to micrometer diameter sizes is possible and thus PIXE allows analysis at high spatial resolution (down to 1 μm). This variant of PIXE, called micro-PIXE (μ-PIXE), is used to determine the distribution of trace elements at the micrometer scale in a wide range of samples. Moreover, owing to its microanalytical capability, the use of PIXE has become an increasingly important tool in the biomedical field, especially in those cases where only small amounts of sample are available to perform the analysis.

Because of its analytical properties, the PIXE technique is now applied in various fields, such as archaeology, art, geology, metallurgy, mineralogy, and so on. But probably biology and medicine are the most important fields of application for this technique. In this section, a brief selection of some applications of PIXE is presented. However, for a comprehensive overview of PIXE applications, reading of a text book on this technique is highly recommended (see Johansson, Akselsson, and Malmqvist, 1995).

PIXE, in principle, can be used to analyze any type of sample. However, as the bombardments are usually done in vacuum, the technique is more suitable for analyzing solids than liquids. With PIXE, most solid samples can be analyzed in their original state (e.g., archaeological samples, soil, hard biological samples, etc.). However, as the penetration of protons in the megaelectron volt range of energy is limited, only elemental information at a depth of several tens of micrometers into the sample is achieved.

- **Biological and medical field.** Most biological samples are composed of an organic matrix and, thus, are well suited to be analyzed by PIXE. The main drawback is that the organic matter is sensitive to radiation damage and to heating effects. These constraints have limited the application of PIXE in some areas of application.

  The preparation of biological/biomedical samples depends on the type of sample, its composition, the aim of the study, and so on. Usually, harder biological and medical samples (nails, hair, bones, etc.) and botanical (leaves, wood, tree rings, etc.) tissues can be analyzed directly by PIXE. In contrast, soft biological and clinical specimens (tissues) require a previous sample treatment, usually based on a cryosectioning of frozen samples, followed by freeze-drying and mounting of a specimen onto a thin plastic foil.

  PIXE applications in the biological and medical field also include the analysis of body fluids (blood, plasma, serum), soft tissues and organs (brain, liver, etc.), human materials (bone, hair, nails, teeth), and so forth. The multielemental character and the high sensitivity of the PIXE method make it also especially suitable for in-situ detection and quantification of metallic ions in single cells or individual proteins after their separation by electrophoresis or to obtain information on metal contents in molecules in combination with high-performance liquid chromatography separations.

  The relatively high spatial lateral resolution of μ-PIXE makes it well-suited also for studying plant material in the botanical and environmental fields of study. For instance, μ-PIXE has been applied as a reliable microanalytical tool for quantitative localization of metals within plant tissues. Although, a limitation of PIXE in this application has been the thermal damage induced to the samples and associated problems of redistribution of elements when performing the analysis, which is inherent when using a charged-particle excitation source.

- **Environmental and earth science applications.** Considering that atmospheric aerosols are often collected as a thin-sample layer onto a thin filter or substrate film, these samples are ideal for analyis by PIXE. The multielemental analysis of airborne particulate material has proven to be one of the most popular applications of PIXE since its development. In addition to atmospheric aerosols, various other types of environmental and geological samples have been analyzed by PIXE. For instance, μ-PIXE offers great potential for studies associated with mineral prospecting and for improving the understanding of basic geological processes. PIXE has also been used in paleontology to study metallic enrichment in fossils.
- **Arts and archaeology.** The PIXE technique has proven to be an extremely rewarding technique for studies related to art and archaeological specimens and artifacts. For instance, an ion-beam analysis facility has been set up in the scientific laboratory of the Louvre Museum in Paris (AGLAE facility of the Centre de Recherche et de Restauration des Musées de France). In this field of study, the nondestructive capability of the technique is more important than in other application and thus the beam size and intensity are selected as a compromise between analytical sensitivity and preventing possible damage to the object under investigation. Both PIXE and μ-PIXE have been used in this field in provenance studies of pottery and earthenware, coins, archeological bronzes, and also to study pigments in paintings for authentication of suspect forgeries or for advancing understanding of ancient painting technologies (μ-PIXE).

7.5 ELECTRON-INDUCED X-RAY EMISSION

In electron-induced X-ray emission techniques or electron-probe X-ray microanalysis (EPXMA), an accelerated and focused electron beam is used for inner shell ionization of the atoms present in a sample. Then, as in XRF and PIXE (see Section 7.4), the characteristic X-rays produced by de-excitation of the vacancies can be measured to obtain information on the elemental composition of the sample. In addition, in some EPXMA techniques, secondary and back-scattered electrons are also monitored to gain information on a sample's surface topography. Within the EPXMA techniques, it is of important to discriminate between scanning electron microscopy (SEM) and electron microprobe analysis (EMPA).

7.5.1 SCANNING ELECTRON MICROSCOPE (SEM)

In a typical SEM, an electron beam that has an energy ranging from 0.2 to 40 keV is emitted from an electron gun and focused by some lenses to give a spot of 0.4–5 nm in diameter on the sample. The primary electron beam interacts with a small sample volume (named the *interaction volume*) which extends from less than 100 nm to about 5 μm into the surface, depending on the sample matrix, electron energy, and sample density. The electron beam is scanned over the surface of the sample in a motion similar to a television camera to produce a rastered digital image. The interaction of the electrons with the sample results in different phenomena (back-scattered electrons, secondary electrons, and X-rays) that can be monitored using an appropriate detection system. The type of detection employed can give different information about the sample. In the most common detection mode, SEM produces very high-resolution images of the morphology (size, shape) and topography of the sample surface. However, to obtain information on the composition of the 3D images of microscopic areas, usually an energy dispersive detection system (EDS) is used to detect the characteristic X-rays produced by the interaction of the beam electrons with the sample atoms. Modern SEM–EDS systems are operated using sophisticated software that allows for mapping of the composition of the elements on the surface of the specimen. For conventional imaging using SEM, specimens must be electrically conductive, at least at their surface, to avoid charging of the sample when scanned by the electron beam. For this reason, usually, a sputtering machine is needed to deposit a thin metal film (gold, graphite, etc.) on nonconductive samples prior to their analysis. A more recent innovation in electron microscopy is environmental SEM (ESEM), which allows samples to be studied at pressures and humidities that approach ambient conditions and even nonconducting specimens may be imaged uncoated.

7.5.2 ELECTRON MICROPROBE ANALYSIS (EMPA)

An electron microprobe analyzer (EMPA) is essentially similar to a SEM, but EMPA instruments are usually equipped with a range of WD detection systems that enable quantitative elemental analyses at high spectral resolution, which makes them generally much more expensive than SEM units. Focusing to micrometer sizes is also possible when working with the EPMA technique and thus one can also get spatially resolved information. The combination of high spatial and spectral resolution is particularly useful for assessing the composition of individual particles or grains and chemical changes on the scale of a few micrometers to millimeters. In practice, this is one of the main applications of EMPA technique. Despite the electron image being rather poor compared to an SEM image, a user can still view the surface with enough magnification to identify features of interest and then analyze the composition of them. Samples to be measured by EMPA must be solid (with a flat surface), conductive, and vacuum compatible. For this reason, usually, a sample pretreatment of the sample is needed before EMPA (e.g., sputtering of the sample surface with a thin metal film, preparation of a polished thin section, etc.).

Usually, EMPA and SEM are highly complementary techniques, especially in the studies related to materials science. EMPA is preferred if quantitative near-surface analysis of small areas is required, while SEM–EDS is selected if high resolution imaging is necessary. Recent developments in instrumentation have focused on combining the advantages of both techniques into a single scanning electron beam instrument. In Table 7.4, the main differences of EMPA and SEM–EDS systems are displayed.

Table 7.4. Main characteristics of EMPA and SEM–EDS systems

Characteristics	EMPA	SEM–EDS
Spectral resolution	High (usually WD detection)	Low (usually ED detection)
Electron image	Rather poor (40000×) 2D images	High magnifications (>100000×) 2D/3D images
Sample information	Morphology: Limited Composition: Good	Topography/Morphology: Good Composition: Limited

WD: Wavelength-dispersive; ED: Energy-dispersive.

CHAPTER 8

Overview of XRF and Related Techniques

8.1 INTRODUCTION

The aim of this chapter is to give a summarized overview of the performance of XRF and related techniques discussed in this book and also to further outline the role of XRF as applied to the chemical analysis field. For such a purpose, a critical comparison of the analytical capabilities of XRF systems with other elemental techniques commonly used for elemental analysis is presented. Trends and future perspectives of XRF are included at the end of this chapter.

8.2 COMPARATIVE PERFORMANCE OF XRF SYSTEMS

Taking into account the instrumental design and analytical capabilities of conventional XRF systems (WDXRF, EDXRF, TXRF) and special configurations/related techniques (μ-XRF, SRXRF, PIXE, EMPA, SEM–EDS) covered in Chapters 5 through 7, the selection of the most appropriate method will be based on the requirements of the given analytical task.

Usually for bulk elemental analysis of solid samples, WDXRF and EDXRF systems are preferred. Using such systems to determine the major, minor, and trace elements is possible for an extensive range of solid samples in different application fields with minimum sample preparation (see Chapter 4). WDXRF is used almost exclusively for routine, high-throughput analyses when there is a great need for fast, accurate analysis, and where flexibility is of little importance. Therefore such systems are widely used, for example, in quality control for industrial routine analyses or to determine the major and minor elements in geological specimens. EDXRF systems lack somewhat in resolution compared with WDXRF systems but have the capability to display information on all elements present in the sample at the same time, which makes EDXRF instrumentation ideal for general troubleshooting problems in many application fields. EDXRF has been particularly effective, inter alia, in the alloy industry, forensic science, and the environmental field. Moreover, the applications of portable EDXRF systems have proven especially useful in art and numismatic studies owing to the possibility to analyze objects that cannot be transported to the laboratory. However, the direct WDXRF or EDXRF analysis of liquid samples is technically difficult and is characterized by particular errors in the results. Liquid samples usually provide a high X-ray scatter background resulting in detection limits in the milligram per liter range which are not satisfactory for most environmental application requirements, but may still be adequate enough for waste water samples. The sensitivity for liquid analysis can be improved using sample preconcentration procedures prior to the WDXRF/EDXRF analysis (see Chapter 4) or by using the TXRF approach. As a consequence of the configuration of TXRF systems, the high background that would generally occur because of scatter from the sample support is absent and, therefore, concentrations in the range of several microgram per liter can be measured in aqueous samples. Taking into account the basic principles and the analytical capabilities of TXRF, since its development, it has been primarily used for chemical micro and trace analyses of small quantities of solutions and suspensions placed on optical flats as sample supports. Moreover, owing to the angle dependence of the fluorescence signal in the range of total reflection, TXRF has also been used to investigate surface impurities and even molecules adsorbed onto flat surfaces. Such examinations were and still are of relevant interest in the semiconductor industry to study possible metal impurities on Si wafers.

If the aim of a study is to not only gain information about elemental composition of a sample but also to selectively analyze small sample areas, then microanalytical techniques, such as μ-XRF, PIXE, or EPXMA, are required instead of the more conventional XRF-configured systems. When using μ-XRF systems, it is possible to record information with an acceptable spatial resolution and sensitivity with a system equipped with a conventional X-ray source, but a significant improvement in terms of spatial resolution can be achieved if accelerated particles (protons and electrons) are used as excitation beam sources (see Table 8.1).

Regarding the sensitivity, as is shown in Table 8.1, relative detection limits with the PIXE technique are typically two orders of magnitude lower than for EMPA owing to the much lower continuum background in the resulting characteristic X-ray spectrum (see Chapter 7 for details). However, the main advantages of EPXMA systems (both SEM and EPMA) over PIXE are a higher speed of analysis, better spatial resolution, better imaging capabilities, and the possibility to determine light elements down to Boron. Moreover EMPA systems are more compact

Table 8.1. Analytical characteristics of micro beam X-ray techniques (modified from Beckhoff et al. (2006))

Excitation beam	**Electrons (15–25 keV)**	**Protons (2–3 MeV)**	**X-rays**
Technique	EPXMA (SEM and EPMA)	PIXE	μ-XRF
Beam size (μm)	0.05	0.3–5	0.5–20
Lateral resolution (μm)	1–2	1–2	0.5–10
Penetration (μm)	1–5	5–20	10–1000
LOD (ppm)	100–10000	1–10	1–100
Sample preparation	Conductive and vacuum compatible	Vacuum compatible	No constraints

and more commonly commercially available. Another interesting characteristic that has to be taken into consideration to select the most appropriate microanalytical technique is the penetration depth of the X-ray beam. Since X-rays penetrate the sample deeper than electrons or protons, this can be advantageous for certain applications. For instance, for the analysis of bulk materials, a deeper penetration of the exciting beam implies that a larger representative sample volume will contribute to the detected signal. In the case of coated materials, it is possible to measure thicker coatings or multiple layers.

Therefore, the selection of the best microanalytical technique is usually based on the lateral resolution, sensitivity, and depth information required for the target application.

8.3 ROLE OF XRF SPECTROMETRY IN ANALYSIS FIELD

As already stated in previous sections of this book, since its development, X-ray methods have been used widely for the determination of major, minor, and trace elements in an extensive range of samples in many diverse different fields of application.

To assess the more recent trends in the role and position of the XRF spectrometric technique within the scientific field, a computer literature search was undertaken using the database *Scopus* to reveal the number of publications in which the word "XRF" appeared in the title, abstract, or key words. The search period covered was from 1990 to 2009. As shown in Figure 8.1, during the last decade, the number of publications on XRF has nearly tripled, mostly as a consequence of technological advances, including the design of low-power microfocus tubes (e.g., in field-portable instruments) and novel X-ray optics and detectors.

However, in addition to XRF, there are other common, robust, and well-established analytical techniques for trace analysis, including atomic absorption spectroscopy (AAS) techniques, such as flame atomic absorption spectroscopy (FAAS) and electrothermal atomic absorption spectroscopy (ETAAS), and inductively coupled plasma techniques (inductively coupled plasma optical emission spectroscopy (ICP-OES) and inductively coupled plasma mass spectrometry (ICP-MS)). Therefore, another computer literature search was carried out using the same

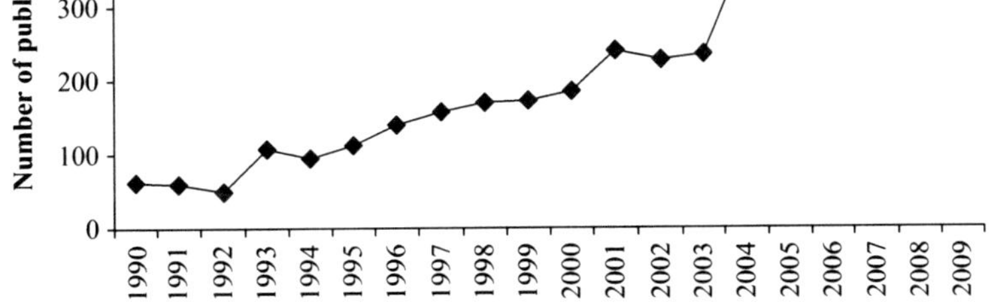

Figure 8.1. Number of articles annually published on XRF in the period 1990–2009.
Source of data: Scopus database.

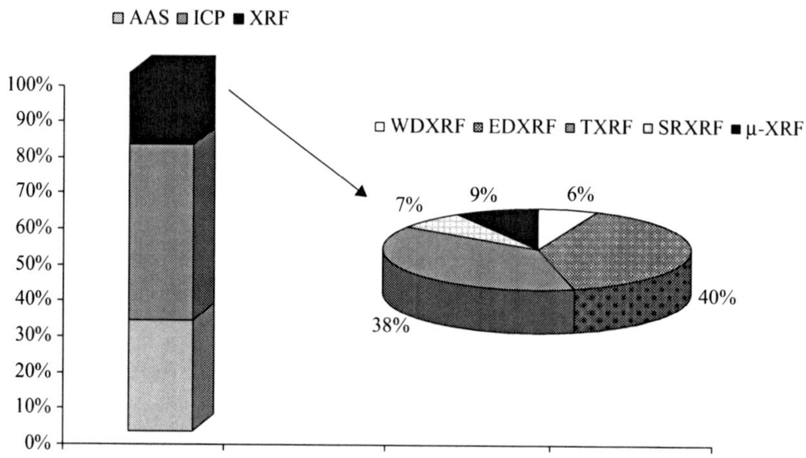

Figure 8.2. Contribution of articles published on XRF, AAS, and ICP in the period 1990–2009.
Source of data: Scopus database.

database covering the same period to assess the number of publication where the word *AAS*, *ICP*, or *XRF* appeared in the title, abstract, or key words. Results obtained are presented in Figure 8.2.

As shown, AAS continues at present to be a very popular, thoroughly tested, and well-established technique in the elemental analysis field. The reasons for the popularity of AAS (including both FAAS and ETAAS) are its relatively low purchase price and maintenance costs, acceptance by many regulatory agencies (e.g., many uses within the US-EPA 7000 analytical method series), detection limits that satisfy the needs of many studies, reasonable precision and accuracy, and simplicity. But AAS also has limitations, such as the sequential character of the determination of the elements, which significantly increases the analysis time, and the impossibility of determining nonmetals such as sulfur. For this reason, ICP techniques are preferred for most applications, as they allow the simultaneous determination of a variety of elements over a wide dynamic concentration range. Moreover, when using ICP-MS systems, detection limits at nanogram per liter level can be achieved.

The main drawback of all the atomic spectroscopic techniques mentioned is that the market for these types of instruments principally offers instrumentation that is initially dedicated to the analysis of liquid samples. Therefore, solid samples have to be brought into a solution form before analysis by means of a previous chemical pretreatment. Sample dissolution is often a tedious and time-consuming step in the analytical procedure that sometimes limits its application in environmental studies where conclusions have to be drawn on the basis of the results of a large number of analyzed samples. For that reason, in many fields, especially those concerned with the analysis of solid samples, XRF is often the technique of choice.

The main drawbacks of XRF instrumentation, restricting their more frequent use for environmental studies, have been their insufficient sensitivity for some important pollutant elements (such as Cd and Pb) and a somewhat poorer precision and accuracy compared to the atomic absorption and emission spectroscopic techniques. Nevertheless, the recent improvements in XRF instrumentation, such as the development of digital signal processing based spectrometers in combination with improved X-ray production using better designs for excitation-detection has added the advantage of increasing instrumental sensitivity, thus allowing the improvement of both precision and productivity. These developments have promoted an increasing interest in using XRF spectroscopy as an alternative to destructive analytical methods in the environmental field.

From Figure 8.2, it can also be deduced that most XRF publications are focused on EDXRF. This clearly reflects the predominant use of more expensive WDXRF systems in routine industrial analysis (where publishing is much less common) while EDXRF is predominantly the technique of choice in academic and research institutions.

Despite the fact that use of TXRF for multielemental chemical analysis is of significance, its relative importance has decreased in recent years mostly owing to the increasing use and popularity of ICP-MS systems. The main reasons for that are its excellent detection limits, the low time required to perform an analysis, and the possibility to get isotopic information with the technique. However, TXRF spectrometry still has some important advantages over ICP-MS systems, including the presence of few spectral interferences, its microanalytical capability (analysis of a microliter or microgram of sample), the possibility of obtaining quantitative results by internal standardization, and also the lower capital investment, running and maintenance costs compared to ICP-MS instruments.

In spite of the lower number of published contributions using SR-XRF and μ-XRF compared to conventional EDXRF systems, mostly owing to their limited availability to the general user's community, the field of SR-XRF and μ-XRF is currently experiencing a significant evolution in instrumentation and it is expected that the role of these XRF variants will increase significantly in the upcoming years, especially as the possibility to obtain spatial resolution information is an important requirement in many fields of application. In Figure 8.3, a useful

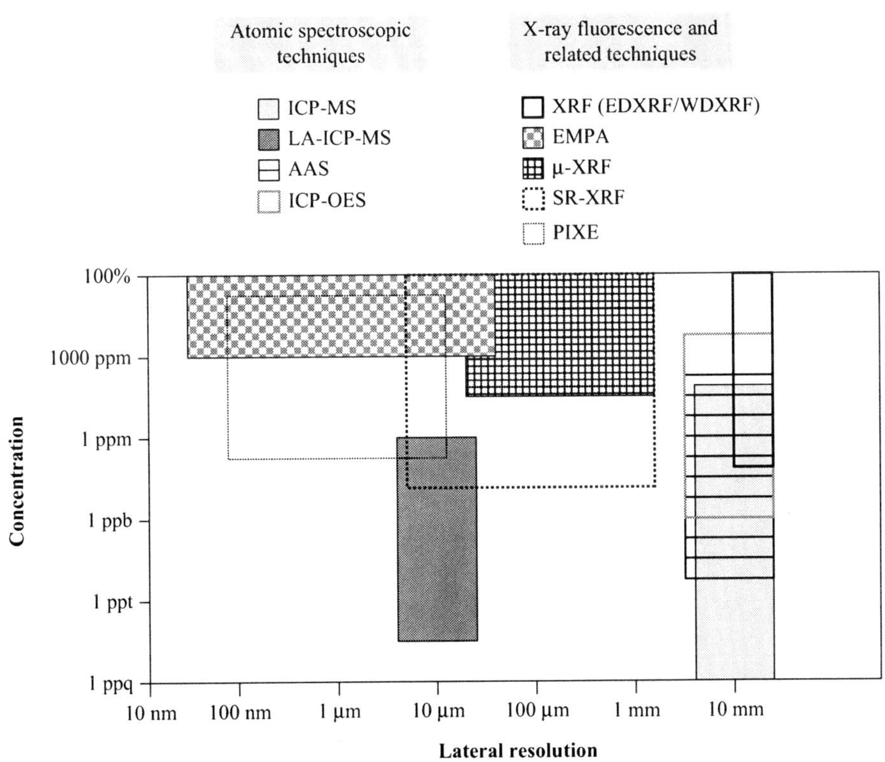

Figure 8.3. Spatial resolution and detection limits of analytical techniques used for elemental analysis (modified from Beckhoff et al. (2006)).

graphic view of the spatial resolution and sensitivity of most analytical techniques used for elemental analysis is displayed. One of the current challenges of analytical techniques is the ability to determine and quantify elements with mass fractions lower than 10^{-10} owing to the requirements imposed by new national and international legislation that limit the allowable concentration of pollutants to ultratrace levels. However, the possibility to get elemental spatial distribution at the micrometer (or even nanometer) scale is also particularly valuable for studying metal composition in heterogeneous samples. As shown in Figure 8.3, most atomic spectroscopic techniques are able to determine elements at trace and ultratrace levels but their spatial resolution is limited. Only when using laser ablation ICP-MS (LA-ICP-MS) can information at the micrometric scale be assessed. Therefore, for spatial resolution analysis, μ-XRF and related techniques are preferred.

To select the best elemental analytical technique for a given purpose, in addition to the limits of detection and spatial resolution capability, other analytical parameters and budgetary considerations are of significance. In Table 8.2, information on this topic is displayed for most analytical techniques used for elemental analysis.

In fact, there is no universal analytical technique fulfilling all the requirements. Each of these techniques has its advantages and drawbacks and thus, few laboratories rely only on one of these analytical methods, but often use complementary combinations of them.

Table 8.2. Analytical characteristics and budgetary consideration for analytical techniques used for elemental analysis

	FAAS	ETAAS	ICP-OES	ICP-MS	XRF
Sample throughput	10–15 s (per element)	3–4 min (per element)	6–60 element (per minute)	2–5 min (all elements)	10–15 min (all elements)
Dynamic range	10^3	10^2	10^5	10^5–10^8	10^4–10^6
Precision	<10%	1%–10 %	1%–5 %	<5%	1%–10%
Interferences					
Spectral	Very few	Very few	Many	Few	Few
Chemical (matrix)	Many	Very many	Very few	Some	Some
Sample amount	mL	μL	mL	mL	mg–g
SemiQ information	No	No	Yes	Yes	Yes
Isotopic analysis	No	No	No	Yes	No
Ease of use	Very easy	Moderate	Easy	Moderate	Easy
Automatization	No	Yes	Yes	Yes	Yes
Capital costs	Low	Medium	High	Very high	Medium/ High
Running costs	Low	Medium	High	Very high	Low
Cost per element analysis	Low	High	Medium	Medium	Low/medium

8.4 FUTURE PERSPECTIVES

Today, XRF is a well-established analytical technique for elemental analysis of a wide variety of samples. In particular, the multielement character, the possibility to perform the analysis directly on solid samples, the possibility to perform qualitative, semi-quantitative, and quantitative determinations, the wide dynamic range, acceptable speed and economy, easy of automation, and versatility are the most important features among the many that have made it a very attractive analytical support tool for both routine quality control in many industries as well as analytical support for the research laboratory. Recent development and commercialization of benchtop/table-top spectrometers, which offer extreme simplicity of operation in a low-cost compact design, have further promoted the approach of this technique in the environmental field for many analytical problems.

Despite the fact that XRF technique is mostly used for the analysis of solid samples, the development and application of new sample treatment/preconcentration procedures in combination with XRF systems will further expand the applications of this technique in the future.

It is expected that future improvements in instrumentation could further increase the analytical sensitivity and spatial resolution and thus, XRF spectrometry will likely offer new possibilities to many fields of application in the coming years. This instrumental progress includes the development of better designs for excitation-detection, new configurations, and also advances in X-ray optics. Of special interest for the future are new third-generation storage rings (used in SR) that are designed to obtain unique intensity and brilliance and thus combine higher spatial resolution with higher sensitivity.

BUYER'S GUIDE TO MANUFACTURERS

A (partial) list of manufacturers (alphabetic order) of XRF spectrometers serving worldwide is provided in this section.

- **Bruker AXS**
  (Web: http://www.bruker-axs.com)
- **CAMECA, S.A**
  (Web: http://www.cameca.fr)
- **Fischer, Ltd**
  (Web: http://www.helmutfischer.com/)
- **Oxford Instruments, Plc**
  (Web: http://www.oxford-instruments.com)
- **PANalytical B.V.**
  (Web: http://www.panalytical.com)
- **Rigaku Corporation**
  (Web: http://www.rigaku.com)
- **Shimadzu Corporation**
  (Web: http://www.shimadzu.com)
- **Spectro Analytical Instruments, GmbH**
  (Web: http://www.spectro.com)
- **Thermo Scientific Corporation**
  (Web: http://www.thermoscientific.com)

Glossary of Abbreviations and Acronyms

- **AFS:** Atomic fluorescence spectrometry
- **APDC:** Ammonium pyrrolidine-dithiocarbamate
- **BFP:** Backscattered fundamental method
- ***Chi* value:** Fit quality value
- **CPA:** Cold plasma ashing
- **DBDTC:** Dibenzyldithiocarbamate
- **DPP:** Digital pulse processor
- **EDXRF:** Energy dispersive X-ray fluorescence
- **EPMA:** Electron probe microanalyzer
- **ESEM:** Environmental scanning electron microscopy
- **ETAAS:** Electrothermal atomic absorption spectroscopy
- **FAAS:** Flame atomic absorption spectroscopy
- **FP:** Fundamental parameter
- **FPC:** Flow proportional counter
- **FPXRF:** Field-portable X-ray fluorescence
- **FWHM:** Full width at half maximum
- **GFAAS:** Graphite furnace atomic absorption spectroscopy
- **GIXRF:** Grazing incidence X-ray fluorescence
- **HE-P-EDXRF:** High-energy polarized-beam EDXRF
- **HPGe:** High purity germanium
- **ICP-AES**: Inductively coupled plasma atomic emission spectroscopy
- **ICP-OES:** Inductively coupled plasma optical emission spectroscopy
- **ICP-MS:** Inductively coupled plasma mass spectrometry
- **INAA:** Instrumental neutron activation analysis
- **IRMM:** Institute of Reference Materials and Measurements
- **LOD:** Limit of detection
- **LOQ:** Limits of quantification
- **MCA:** Multichannel analyzer
- **MIBK:** Methylisobutylketone
- **Micro-XRF** (also μ-XRF)**:** Microbeam XRF
- **NIST:** National Institute of Standards and Technology

- **P-EDXRF:** Polarized-beam energy dispersive X-ray fluorescence
- **PIXE:** Particle induced X-ray emission
- **RSD:** Relative standard deviation
- **SC:** Scintillator counter
- **SDD:** Silicon drift detector
- **SEM:** Scanning electron microscopy
- **SRXRF:** Synchrotron radiation X-ray fluorescence
- **TXRF:** Total reflection X-ray fluorescence
- **VPD:** Vapor-phase decomposition
- **WDXRF:** Wavelength dispersive X-ray fluorescence
- **XAS:** X-ray absorption spectroscopy
- **XRD:** X-ray diffraction
- **Z:** Atomic number
- **Z_{eff}:** Effective atomic number

References

Aiginger, H., and P. Wobrauschek. 1974. "A Method for Quantitative X-Ray Fluorescence Analysis in the Nanogram Region." *Nuclear Instruments and Methods 114*: 157. doi: http://dx.doi.org/10.1016/0029-554X(74)90352-8.

Currie, L. A. 1995. "Nomenclature in Evaluation of Analytical Methods Including Detection Limits and Quantification Capabilities." *Pure and Applied Chemistry 67*: 1699–723. doi: http://dx.doi.org/10.1351/pac199567101699

Eksperiandova, L. P., Y. N. Makarovska, and A. B. Blank. 1998. "Determination of Small Quantities of Heavy Metals in Water-Soluble Salts and Natural Water by X-Ray Fluorescence." *Analytica Chimica Acta 371*: 105–8. doi: http://dx.doi.org/10.1016/S0003-2670(98)00289-X

Fittschen, U. E. A., and G. Falkenberg. 2011. "Trends in Environmental Science Using Microscopic X-Ray Fluorescence." *Spectrochimica Acta Part B 66*: 567–80. doi: http://dx.doi.org/10.1016/j.sab.2011.06.006

Fontàs, C., E. Marguí, M. Hidalgo, and I. Queralt. 2009. "Improvement Approaches for the Determination of Cr(VI), Cd(II), Pd(II) and Pt(IV) Contained in Aqueous Samples by Conventional XRF Instrumentation." *X-Ray Spectrometry 38*: 9–17. doi: http://dx.doi.org/10.1002/xrs.1093

Gonzalez-Fernandez, O., M. Hidalgo, E. Marguí, M. L. Carvalho, and I. Queralt. 2008. "Heavy Metals' Content of Automotive Shredder Residues (ASR): Evaluation of Environmental Risk." *Environmental Pollution 153*: 476–82. doi: http://dx.doi.org/10.1016/j.envpol.2007.08.002

Gonzalez-Fernandez, O., A. M. Jurado-Roldan, and I. Queralt. 2011. "Geochemical and Mineralogical Features of Overbank and Stream Sediments of the Beal Wadi (Cartagena-La Union Mining District, SE Spain): Relation to Former Lead-Zinc Mining Activities and its Environmental Risk." *Water, Air and Soil pollution 215*: 55–65. doi: http://dx.doi.org/10.1007/s11270-010-0458-1

Gonzalez-Fernandez, M., E. Marguí, and I. Queralt. 2009. Multielement Analysis of Dried Residue from Metal-Bearing Waters by Wavelength Dispersive X-Ray Fluorescence Spectrometry." *Spectrochimica Acta Part B 64*: 184–90. doi: http://dx.doi.org/10.1016/j.sab.2008.11.011

Horntrich, C., P. Kregsamer, P. Wobrauschek, and C. Streli. 2012. "Production of the Ideal Sample Shape for Total Reflection X-Ray Fluorescence Analysis." *Spectrochimica Acta Part B 77*: 31–4. doi: http://dx.doi.org/10.1016/j.sab.2012.07.026

Janssens, K., W. De Nolf, G. Van Der Snickt, L. Vincze, B. Vekemans, R. Terzano, and F. E. Brenker. 2010. "Recent Trends in Quantitative Aspects of Microscopic X-Ray Fluorescence Analysis." *Trends in Analytical Chemistry 29*: 464–78. doi: http://dx.doi.org/10.1016/j.trac.2010.03.003

Koulouridakis, P. E., and N. G. Kallithrakas-Kontos. 2004. Selective Mercury Determination After Membrane Complexation and Total Reflection X-Ray Fluorescence Analysis." *Analytical Chemistry 76*: 4315–19. doi: http://dx.doi.org/10.1021/ac049780a

Marguí, E., G. H. Floor, M. Hidalgo, P. Kregsamer, G. Roman-Ross, C. Streli, and I. Queralt. 2010a. Analytical Possibilities of Total Reflection X-Ray Spectrometry (TXRF) for Trace Selenium

Determination in Soils." *Analytical Chemistry 82*: 7744–81. doi: http://dx.doi.org/10.1021/ac101615w

Marguí, E., R. Van Grieken, C. Fontas, M. Hidalgo, and I. Queralt. 2010b. "Preconcentration Methods for the Analysis of Liquid Samples by X-Ray Fluorescence Techniques." *Applied Spectroscopy Reviews 45*: 179–205. doi: http://dx.doi.org/10.1080/05704920903584198

Marguí, E., J. C. Tapias, A. Casas, M. Hidalgo, and I. Queralt. 2010c. "Analysis of Inlet and Outlet Industrial Wastewater Effluents by Means of Benchtop Total Reflection X-Ray Fluorescence Spectrometry." *Chemosphere 80*: 263–70. doi: http://dx.doi.org/10.1016/j.chemosphere.2010.04.027

Marguí, E., C. Fontàs, K. Van Meel, M. Hidalgo, and I. Queralt. 2008. "XRF Trace Metal Analysis of Environmental Liquid Samples After Membrane Preconcentration." *Spectroscopy Europe 20*: 18–21.

Marguí, E., K. Van Meel, R. E. Van Grieken, A. Buendia, C. Fontàs, M. Hidalgo, and I. Queralt. 2009a. "Method for the Determination of Pd-Catalysts Residues in Active Pharmaceutical Ingredients by Means of High-Energy Polarized-Beam Energy Dispersive X-Ray Fluorescence." *Analytical Chemistry 81*: 1404–10. doi: http://dx.doi.org/10.1021/ac8021373

Marguí, E., C. Fontàs, A. Buendia, M. Hidalgo, and I. Queralt. 2009b. "Determination of Metal Residues in Active Pharmaceutical Ingredients According to European Current Legislation by Using X-Ray Fluorescence Spectrometry." *Journal of Analytical Atomic Spectrometry 24*: 1253–7. doi: http://dx.doi.org/10.1039/b904064a

Marguí, E., I. Queralt, M. L. Carvalho, and M. Hidalgo. 2005a. "Comparison of EDXRF and ICP-OES After Microwave Digestion for Element Determination in Plant Specimens from an Abandoned Mining Area." *Analytica Chimica Acta 549*: 197–204. doi: http://dx.doi.org/10.1016/j.aca.2005.06.035

Marguí, E., M. Hidalgo, and I. Queralt. 2005b. Multielemental Fast Analysis of Vegetation Samples by Wavelength Dispersive X-Ray Fluorescence Spectrometry: Possibilities and Drawbacks." *Spectrochimica Acta Part B 60*: 1363–72. doi: http://dx.doi.org/10.1016/j.sab.2005.08.004

Marguí, E., M. Hidalgo, I. Queralt, K. Van Meel, and C. Fontas. 2012. "Analytical Possibilities of Laboratory, Benchtop and Handheld X-Ray Fluorescence Systems for Detection of Metals in Aqueous Samples Pre-Concentrated with Solid-Phase Extraction Disks." *Spectrochimica Acta Part B 67*: 17–23. doi: http://dx.doi.org/10.1016/j.sab.2011.12.004

Marguí, E., R. Padilla, M. Hidalgo, I. Queralt, and R. Van Grieken. 2006. "High-Energy Polarised-Beam EDXRF for Trace Metal Analysis of Vegetation Samples in Environmental Studies." *X-Ray Spectrometry 35*: 169–77. doi: http://dx.doi.org/10.1002/xrs.890

Meirer, F., A. Singh, G. Pepponi, C. Streli, T. Homma, and P. Pianetta. 2010. "Synchrotron Radiation-Induced Total Reflection X-Ray Fluorescence Analysis." *Trends in Analytical Chemistry 29*: 479–96. doi: http://dx.doi.org/10.1016/j.trac.2010.04.001

Miller, T. C., E. P. Hastings, and G. J. Havrilla. 2006. "Automated Printing Technology as a New Tool for Liquid Sample Preparation for Micro X-Ray Fluorescence (MXRF)." *X-Ray Spectrometry* 35 (2): 131–6. doi: http://dx.doi.org/10.1002/xrs.882

Miller, T. C., and G. J. Havrilla. 2004. "Nanodroplets: A New Method for Dried Spot Preparation and Analysis." *X-Ray Spectrometry* 33 (2): 101–6. doi: http://dx.doi.org/10.1002/xrs.710

Padilla Álvarez, R., A. Markowicz, D. Wegrzynek, E. Chinea-Cano, S. A. Bamford, and D. Hernández Torres. 2007. "Quality Management and Method Validation in EDXRF Analysis." *X-Ray Spectrometry 36*: 27–34. doi: http://dx.doi.org/10.1002/xrs.923

Pahlke, S. 2003. "Quo Vadis Total Reflection X-Ray Fluorescence." *Spectrochimica Acta Part B 58*: 2025–38. doi: http://dx.doi.org/10.1016/S0584-8547(03)00193-9

Parsons, C., E. Marguí, E. Pili, G. H. Floor, G. Roman-Ross, and L. Charlet. (in press)- "Quantification of trace arsenic in soils by field-portable X-ray fluorescence spectrometry: Considerations for sample preparation and measurement conditions". *Journal of Hazardous Materials.* doi: http://dx.doi.org/10.1016/j.jhazmat.2012.07.001

Pérez, R. D., M. Rubio, C. A. Pérez, A. R. Eynard, and G. A. Bongiovanni. 2006. "Study of the Effects of Chronic Arsenic Poisoning in Rat Kidney by Means of Synchrotron Microscopic X-Ray Fluorescence Analysis." *X-ray Spectrometry 35*: 352–8. doi: http://dx.doi.org/10.1002/xrs.918

Prange, A., and H. Schwenke. 1989. "Sample Treatments for TXRF-Requirements and Prospects." *Advances in X-Ray Analysis 32*: 211–18.

Sitko, R., and Zawisza, B. 2012. Quantification in X-ray fluorescence spectrometry, in X-ray Spectroscopy, S.K. Sharma (Ed.), InTech, pp.137–162.

Rousseau, R. M. 2001. "Detection Limit and Estimate of Uncertainty of Analytical Results." *The Rigaku Journal 18*: 33–47.

Somogyi, A., M. Drakopoulos, L. Vincze, B. Vekemans, C. Camerani, K. Janssens, A. Snigirev, and F. Adams. 2001. "ID18F: A New Micro-X-Ray Fluorescence End-Station at the European Synchrotron Radiation Facility (ESRF): Preliminary Results." *X-Ray Spectromery 30*: 242–52. doi: http://dx.doi.org/10.1002/xrs.494

Spolnik, Z., K. Belikov, K. Van Meel, E. Adriaenssens, F. De Roeck, R. Van Grieken. 2005. "Optimization of Measurement Conditions of an Energy Dispersive X-Ray Fluorescence Spectrometer with High-Energy Polarized Beam Excitation for Analysis of Aerosols Filters." *Applied Spectroscopy 59*: 1465–9. doi: http://dx.doi.org/10.1366/0003702057751442647

Streli, C. 2006. "Recent Advances in TXRF." *Applied Spectroscopy Reviews 41*: 473–89. doi: http://dx.doi.org/10.1080/10543400600809318

Van Grieken, R. 1982. "Preconcentration Methods for the Analysis of Water by X-Ray Spectrometric Techniques." *Analytica Chimica Acta 143*: 3–34. doi: http://dx.doi.org/10.1016/S0003-2670(01)95486-8

Von Bohlen, A. 2004. "Total Reflection X-Ray Fluorescence Spectrometry – A Versatile Tool for Ultra-Micro Analysis of Objects of Cultural Heritage." *e-Preservation Science 1*: 23–34 (www.e-PreservationScience.org).

Wang, J., M. K. Balazs, P. Pianetta, K. Baur, and S. Brennan. 2000. "Analytical Techniques for Trace Elemental Analyses on Wafer Surfaces for Monitoring and Controlling Contamination." *Stanford Publications*. pp. 1–9.

Willis, J. P., and G. R. Lachance. 2004. "Comparison Between Some Common Influence Coefficient Algorithms." *X-Ray Spectrometry 33*: 181–8. doi: http://dx.doi.org/10.1002/xrs.681

Woelfl, S., M. Mages, and F. Encina. 2003. "Cold Plasma Ashing Improves the Trace Element Detection of Single Daphnia Specimens by Total Reflection X-Ray Fluorescence Spectrometry." *Spectrochimica Acta Part B 58*: 2157–68. doi: http://dx.doi.org/10.1016/S0584-8547(03)00196-4

Yoneda, Y., and T. Horiuchi. 1971. "Optical Flats for Use in X-Ray Spectrochemical Microanalyis." *Review of Scientific Instruments 42*: 1069. doi: http://dx.doi.org/10.1063/1.1685282

BIBLIOGRAPHY

BOOKS AND ENCYCLOPEDIA CHAPTERS

Beckhoff, B., B. Kanngiesser, N. Langhoff, R. Wedell, and H. Wolff. 2006. *Practical X-ray Fluorescence Analysis*. Berlin: Springer-Verlag.

Brouwer, P. 2010. *Theory of XRF. Getting Acquainted with the Principles*. Almelo, The Netherlands: PANalytical BV.

Buhrke, V. E., R. Jenkins, and D. K. Smith. 1998. *A Practical Guide for the Preparation of Specimens for X-Ray Fluorescence and X-Ray Diffraction Analysis*. New York: John Wiley & Sons.

Janssens, K. H. A., F. Adams, and A. Rindby. 2000. *Microscopic X-Ray Fluorescence Analysis*. Chichester, England: John Wiley & Sons, Ltd.

Jenkins, R. 1996. *X-Ray Fluorescence Spectrometry*. Chemical analysis. A series of monograph on analytical chemistry and its applications. Volume 152. Chichester, England: John Wiley & Sons.

Johansson, T. B., K. R. Akselsson, and K. G. Malmqvist. 1995. *Particle-Induced X-Ray Emission Spectrometry (PIXE)*. Chichester, England: John Wiley & sons.

Klockenkämper, R. 1997. *Total Reflection X-Ray Fluorescence Analysis*. A series of monographs on analytical chemistry and its applications. Volume 152. Chichester, England: John Willey & Sons.

Lachance, E. R., and F. Claisse. 1995. *Quantitative X-Ray Fluorescence Analysis: Theory and Application*. Chichester, England: John Wiley & Sons.

Marguí, E., I. Queralt, and R. Van Grieken. 2009. "X-Ray Fluorescence Analysis, Sample Preparation for." In *Encyclopedia of Analytical Chemistry*, edited by R. A. Meyers. Chichester, England: John Wiley & Sons Ltd.

Potts, P. J., and M. West. 2008. *Portable X-Ray Fluorescence Spectrometry. Capabilities for In-situ Analysis*. Cambridge: RSC Publishing.

Shackley, S. M. 2011. *X-Ray Fluorescence Spectrometry in Geoarchaeology*. New York: Springer.

Tsuji, K., J. Injuk, and R. Van Grieken. 2004. X-*Ray Spectrometry: Recent Technological Advances*. Chichester, England: John Wiley & Sons Ltd.

Van Grieken, R.E., and A. A. Markowicz. 2002. *Handbook of X-Ray Spectrometry*, Second Edition. Revised and Expanded. New York: Marcel Dekker, Inc.

JOURNALS

Most of the research articles related to the development and application of XRF methods are published in scientific journals specializing in the topics of Spectroscopy or Analytical Chemistry disciplines (see following list). In this context, the annual review published by the Royal Society of Chemistry "Atomic Spectrometry update—X-ray fluorescence Spectrometry" (*Journal of Analytical Atomic Spectrometry*,

RSC Publishing, Cambridge) offers the reader a wealth of information published annually covering analytical endeavors using conventional and hyphenated XRF techniques, and the biannual reviews on "X-Ray Spectrometry" of Analytical Chemistry (ACS Publications, Washington) are also recommended.

- **Spectroscopic journals:**
  - Journal of Analytical Atomic Spectroscopy
  - Spectrochimica Acta Part B: Atomic Spectroscopy
  - X-Ray Spectrometry
  - Advances in X-Ray Analysis
  - Applied Spectroscopy
  - Spectroscopy
  - Applied Spectroscopy Reviews (Only review contributions)

- **Analytical Chemistry journals:**
  - Analytical Chemistry
  - Analytica Chimica Acta
  - Analytical and Bioanalytical Chemistry
  - Microchimica Acta
  - Talanta
  - The Analyst
  - Analytical Letters
  - Analytical Sciences
  - TrAC Trends in Analytical Chemistry (Only review contributions)

However, it is interesting to note that instrumental developments related to the XRF spectrometry are more usually published in scientific journals focused on the disciplines of physics and instrument development:

- Nuclear Instrument & Methods in Physics Research. Section A: Accelerators, Spectrometers, Detectors and Associated Equipment
- Nuclear Instrument & Methods in Physics Research. Section B: Beam Interactions with Materials and Atoms
- Nuclear Engineering and Technology
- Review of Scientific Instruments
- Journal of Radioanalytical and Nuclear Chemistry
- Physical Chemistry Chemical Physics
- Chemical Physics Letters
- Physical Review Letters

INDEX

CPSIA information can be obtained at www.ICGtesting.com
Printed in the USA
BVOW01*0246071013

332647BV00003B/3/P

9 781606 503911